W9-CVO-253

FIBONACCI & LUCAS NUMBERS, AND THE GOLDEN SECTION
Theory and Applications

Mario Merz *Fibonacci tables* Tate Gallery, London.

FIBONACCI & LUCAS NUMBERS, AND THE GOLDEN SECTION

Theory and Applications

S. VAJDA, Dr. Phil., D. Tech. h.c.
Department of Mathematics, University of Sussex

ELLIS HORWOOD LIMITED
Publishers · Chichester

Halsted Press: a division of
JOHN WILEY & SONS
New York · Chichester · Brisbane · Toronto

First published in 1989 by
ELLIS HORWOOD LIMITED
Market Cross House, Cooper Street,
Chichester, West Sussex, PO19 1EB, England
The publisher's colophon is reproduced from James Gillison's drawing of the ancient Market Cross, Chichester.

Distributors:
Australia and New Zealand:
JACARANDA WILEY LIMITED
GPO Box 859, Brisbane, Queensland 4001, Australia
Canada:
JOHN WILEY & SONS CANADA LIMITED
22 Worcester Road, Rexdale, Ontario, Canada
Europe and Africa:
JOHN WILEY & SONS LIMITED
Baffins Lane, Chichester, West Sussex, England
North and South America and the rest of the world:
Halsted Press: a division of
JOHN WILEY & SONS
605 Third Avenue, New York, NY 10158, USA
South-East Asia
JOHN WILEY & SONS (SEA) PTE LIMITED
37 Jalan Pemimpin # 05–04
Block B, Union Industrial Building, Singapore 2057
Indian Subcontinent
WILEY EASTERN LIMITED
4835/24 Ansari Road
Daryaganj, New Delhi 110002, India

British Library Cataloguing in Publication Data
Vajda, Steven, *1901–*
Fibonacci & Lucas numbers and the golden section
1. Fibonacci numbers
I. Title II. Series
512′.72

Library of Congress data available

ISBN 0–7458–0715–1 (Ellis Horwood Limited)
ISBN 0–470–21508–9 (Halsted Press)

Typeset in Times by Ellis Horwood Limited
Printed in Great Britain by Hartnolls, Bodmin

Table of contents

Preface

Fibonacci numbers have attracted the attention of professional as well as amateur mathematicians. Their deceptively simple definition implies a large variety of relationships. They appear, often surprisingly, as answers to intricate problems, in conventional and in recreational Mathematics. Moreover, they are relatives of the Golden Section, which itself appears in the study of Nature, and of Art.

The theoretical context of the present book fits into algebra, analysis, geometry, probability theory, computational aspects and, of course, the combinatorial side of number theory.

The Golden Section was studied in antiquity, while Fibonacci numbers are less than 800 years old, and their serious study dates back only about 150 years. An exhaustive review of the development of our subject is contained in L. E. Dickson's *History of the theory of numbers*.

Our book starts with a brief survey of problems which are solved by Fibonacci numbers and paints their wider background in Chapter II. While the next chapters exhibit mainly formal relationships, the opportunity is taken for sidelines on more light-hearted matters.

In Chapter VI the stress is on divisibility properties, which leads to features connected with the generation of random numbers, much used in computational simulation. Chapter IX on Fibonacci Representation serves, apart from its intrinsic interest, as an introduction to the theory and practice of some entertaining games.

In Chapter XII a letter from Professor B. W. Conolly is quoted about meta-Fibonacci numbers, a largely unexplored territory which offers, so far, only glimpses into an intriguing field. The next two chapters, XIII and XIV, deal with the Golden Section in the plane and in space. The platonic solids are described and some of their less familar features are exhibited.

The Appendix describes briefly some theory which is assumed to be known when reading the main text. A list of formulae, a table of Fibonacci numbers and of Lucas numbers up to order 30, a bibliography, and an index complete the presentation.

A number of theorems are mentioned throughout the book which were first communicated in *The Fibonacci Quarterly*, a 'journal devoted to the study of integers with special properties'. Relatively recent results are combined with earlier material, probably the first time in a comprehensive volume.

I have had stimulating discussions with many colleagues, and I must mention with special gratitude B. W. Conolly, who has himself contributed to the exploration of a fascinating subject.

The figures were prepared by Neil R. Firth, and the frontispiece is a photograph, reproduced by permission, of a picture in the Tate Gallery, London, by M. Merz.

January 1989 S. Vajda

I

Introduction

1. Leonardo of Pisa (c 1170–1250), called Fibonacci, that is son of Bonaccio, has been hailed as the greatest European mathematician of the Middle Ages. He grew up in North Africa, where he became acquainted with the advanced mathematical knowledge of Arabic scholars. In his *Liber Abaci* (1202, revised edition 1228) he advocated the use of those Arabic numerals which we use to-day.

In the same book, he posed the following problem:

A pair of newly born rabbits is brought into a confined place. This pair, and every later pair, begets one new pair every other month, starting in their second month of age. How many pairs will there be after one, two, ... months, assuming that no deaths occur?

In their catalogue information on Merz's picture *Fibonacci Tables* (our frontispiece) the Tate Gallery states, rather quaintly, that the series was 'originally applied to the understanding of reproduction in rabbits.'

At the start of the first month we have that ancestor pair. At the start of the second month, there will still be only that pair, but at the start of the third month another pair has arrived. Generally, let there be F_n pairs at the start of the nth month, and F_{n+1} at the start of the $(n+1)$th month. The latter will still be there at the beginning of the $(n+2)$th month, but the total number of rabbit pairs will have increased by the offspring of all those who were there at the start of the nth month. Hence

$$F_{n+2} = F_{n+1} + F_n \ . \qquad (1)$$

We call these numbers Fibonacci numbers. Their sequence starts, if we let n also be 0, as follows:

n	0	1	2	3	4	5	.	.
F_n	0	1	1	2	3	5	.	.

There is a more extensive listing at the end of this book.

2. The rule (1) is characteristic of the Fibonacci sequence, but the latter is only defined when the first two values, the 'seed' (or indeed any two values of it) are given.

Rule (1) can be used to extend the sequence backwards, thus

$$F_{-1} = F_1 - F_0 \,, \qquad F_{-2} = F_0 - F_{-1} \,, \qquad \text{and so on.}$$

This produces

n	0	1	2	3	4	5	. . .
F_n	0	1	−1	2	−3	5	. . .

and generally

$$F_{-n} = (-1)^{n+1} F_n \,. \tag{2}$$

We shall call any sequence which follows rule (1) a generalized Fibonacci sequence defined by

$$G_{n+2} = G_{n+1} + G_n \,. \tag{3}$$

(see Horadam, 1961).

For instance, let the seed be (2, 1). The numbers thus generated will be called Lucas numbers, denoted by L_n. We have

n	0	1	2	3	4	5 . . .
L_n	2	1	3	4	7	11 . . .

and, extending backwards,

n	0	1	2	3	4	5	. . .
L_{-n}	2	−1	3	−4	7	−11	. . .

Generally,

$$L_{-n} = (-1)^n L_n \,. \tag{4}$$

3. Let us now look at a problem which is structurally similar to Fibonacci's rabbit problem (Tucker, 1980, pp. 112–113).

A staircase has n steps. We climb it by taking one, or by taking two steps at a time. How many ways (say S_n) are there to climb the staircase?

When we start to climb, we can take one step, or two steps to begin with. If we take one step, then there are S_{n-1} possibilities to continue climbing the remaining $n-1$ steps. If we take two steps, then the remaining possibilities are S_{n-2} in number.

Hence

$$S_n = S_{n-1} + S_{n-2} .$$

This formula is equivalent to (1), but we must still determine S_1 and S_2. Now when the staircase consists of one single step, then there is only one possibility of climbing it. $S_1 = 1$. If there are two steps, then there are two possibilities: take two steps, or take two single steps in succession: $S_2 = 2$. So we obtain the sequence

n	1	2	3	4	5	. . .
S_n	1	2	3	5	8	. . .

This is again the Fibonacci sequence, but shifted by one term:

$$S_n = F_{n+1} .$$

4. Consider generations of honeybees. Eggs of workers develop without fertilization into drones, while the queen's eggs are fertilized by drones and develop into females, either queens or workers. Thus a drone has only a single parent, a female, while a female has two parents, one male and one female.

Call the parent of a drone 'generation 1', its grandparents 'generation 2', and so on. In generation 1, we have one female, written $f_1 = 1$, and no male, $m_1 = 0$. In other generations, the following holds:

$$m_{n+1} = f_n \qquad f_{n+1} = f_n + m_n$$
$$m_{n+2} = f_{n+1} \qquad f_{n+2} = f_{n+1} + m_{n+1} .$$

Hence

$$m_{n+2} = m_{n+1} + m_n \quad \text{and} \quad f_{n+2} = f_{n+1} + f_n .$$

Both the sequences of the f_n and those of the m_n are Fibonacci sequences, but with different seeds. In fact

	n	1	2	3	4	5	. . .	
	m_n	0	1	1	2	3	. . .	i.e. $m_n = F_{n-1}$
	f_n	1	1	2	3	5	. . .	i.e. $f_n = F_n$
adding		1	2	3	5	8	. . .	i.e. $m_n + f_n = F_{n-1} + F_n = F_{nf1}$

in generation n, there are F_{n+1} ancestors.

5. Consider the natural numbers 1, 2, 3, . . . , n, in this order. Find the number of those permutations where each number either stays fixed, or changes place with one of its neighbours. More than one of such changes can take place at the same time.

Denote the number of such permuations of n number by g_n. If n stays fixed, then there are g_{n-1} possible permutations. If n and $n-1$ are exchanged, then there are g_{n-2} possibilities left. Hence

$$g_n = g_{n-1} + g_{n-2} \; .$$

To determine g_n completely, consider g_1, which is obviously 1, and $g_2 = 2$ (either leave 1 and 2 in their position, or exchange them). The sequence of the g_n is 1 2 3 5 8 . . . , that is $g_n = F_{n+1}$.

Now consider the same problem, but with the numbers arranged round a circle. Denote the number of possible arrangements as described above by h_n. Assume $n \geqslant 3$.

If 1 and n interchange, then there are g_{n-2} possibilities left. If they do not interchange, then we have precisely the previous problem, with g_n possibilities. Therefore

$$h_n = g_n + g_{n-2} = F_{n+1} + F_{n-1} \; .$$

This time we obtain the generalized Fibonacci sequence

n	3	4	5	. . .
h_n	4	7	11	. . .

In fact, $h_n = L_n$.

The last two examples were communicated by E. K. Lloyd (1985, pp. 553, 564).

6. In how many ways is it possible to obtain n as a sum of positive integers larger than 1?

Let us call this number T_n. For instance, $T_7 = 8$; 7 equals

$$2+2+3 \; , \quad 2+3+2 \; , \quad 2+5 \; , \quad 3+2+2 \; ,$$
$$3+4 \; , \quad 4+3 \; , \quad 5+2 \; , \quad 7 \; .$$

Three of these sums start with 2. If, in these sums, we omit 2, then we have all the sums making up 5. Hence $T_5 = 3$. Five of the sums start with a number larger than 2. If we reduce the first term in these sums by 1, then we obtain all those sums which make up 6. Hence $T_6 = 5$.

Applying this procedure to any n, then we find $T_{n+2} = T_{n+1} + T_n$. Since $T_2 = 1$ and $T_3 = 1$, we conclude that $T_n = F_{n-1}$ (Netto, 1901, p. 134). In a similar, though more general, case, we ask the question: in how many ways is it possible to obtain n as a sum of the integers 1, 2, . . . , m?

We call this number $T_n^{(m)}$. For instance, $T_4^{(3)} = 7$. 4 equals

$$1+1+1+1\ ,\quad 1+1+2\ ,\quad 1+2+1\ ,\quad 1+3\ ,$$
$$2+1+1\ ,\quad 2+2\ ,\quad 3+1\ .$$

Observing the first term in these sums, and the result of omitting it, we find

$$\mathbf{T}_4^{(3)} = T_3^{(3)} + T_2^{(3)} + T_1^{(3)}\ .$$

Now

$$T_1^{(3)} = 1\ ,\qquad T_2^{(3)} = 2\ ,\qquad T_3^{(3)} = 4 \qquad \text{and } 1+2+4 = 7\ .$$

Generally,

$$T_{n+1}^{(m)} = T_1^{(m)} + T_2^{(m)} + T_3^{(m)} + \ldots + T_n^{(m)}\ .$$

This is the mechanism for a recurrence of order n. (See Chapter II.) (Netto, 1901, 136–137).

7. We mention now two cases concerning probabilities.

(a) A fair coin is tossed until two consecutive heads have appeared. The probability of the sequence terminating after n tosses is $F_{n-1}/2^n$.

(b) A fair coin is tossed until either three consecutive heads or three consecutive tails have appeared. The probability of the sequence terminating after n tosses is $F_{n-2}/2^{n-1}$.

These two statements will be proved in Chapter IV, section 3.

8. Find the number of possible sequences of n digits, each digit being either 0 or 1, without two 0s being consecutive anywhere in the sequence.

Let this number be $P(n)$. There are $P(n-1)$ ways, starting with 1, and $P(n-2)$ ways, starting with 0 1. Thus

$$P(n) = P(n-1) + P(n-2)\ .$$

If $n = 1$, then there are two possibilities, viz. 0 or 1. If $n = 2$, then there are three possibilities, viz. 1 1, or 0 1, or 1 0. Hence $P(1) = 2$, $P(2) = 3$, and generally $P(n) = F_{n+2}$.

Actually, this result follows from the answer to the problem of section 7(a). Consider the 2^n possible sequences of 0s and 1s, with equal probabilities for either to appear in the mth place ($m = 1, 2, \ldots, n$). The probability of two 0s having appeared after two steps is $F_1/2^2$, that of two 0s having appeared the first time after three steps, that is 1 0 0, is $F_2/2^3$, after n steps $F_{n-1}/2^n$. Hence the probability of two consecutive 0s having appeared the first time after i steps is $F_{i-1}/2^i$, and that of no such pairs of

0s having appeared at all during the first n steps is

$$1-\sum_{i=2}^{n}\frac{F_{i-1}}{2^i}$$

This equals indeed $F_{n+2}/2^n$ (see formula (37a) in Chapter II).

9. Now imagine that we have a string of m ($\geqslant 2$) digits, each 0 or 1, this string being periodically repeated. Suppose again that no two 0s must appear consecutively. How many strings satisfy this condition? Let the number be $T(m)$.

This problem differs from problem of section 8 in that now the string must not finish with 0, if it started with 0. If, then, the string of m digits starts with 1, then we have $P(m-1)$ possibilities, where $P(m)$ is defined in section 8. But if the string starts with 0, then neither the second, nor the mth digit can be 0, so that there are $P(m-3)$ possibilities left.

To make the situation quite clear, we quote the example for $m = 4$. Then $P(4) =$ 7, as follows:

01010101 . . . , 01110111 . . . , 10111011 . . . , 11011101 . . . ,
11111111 . . . , 10101010 . . . , 11101110 . . . .

(Observe that, for instance, the string 0101 repeated is different from the string 1010 repeated.)

We have seen that $P(m) = F_{m+2}$. Therefore

$$T(m) = P(m-1)+P(m-3) = F_{m+1}+F_{m-1} = L_m \; .$$

The two problems of sections 8 and 9 are mentioned, in a different form, in Rényi (1984, pp. 87–91).

10. Huntley (1970, p. 155) mentions that F_{n+2} is the number of possible paths of a ray reflected n times when passing through two panes of glass, and Gardner (1968, p. 162) mentions that Leo Moser has studied these paths. Example $n = 3$, $F_5 = 5$.

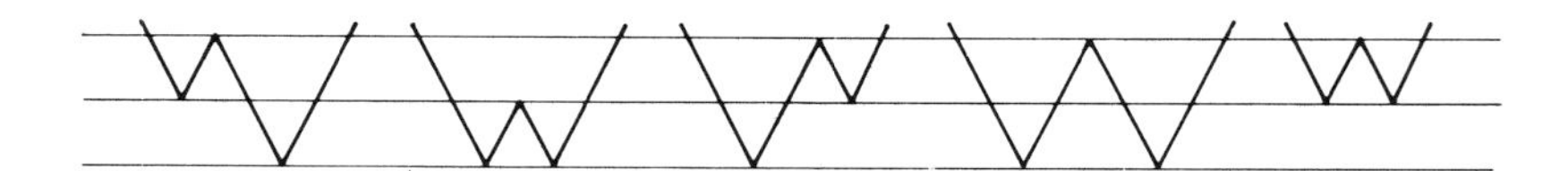

11. In graph theory, an n-gon ($n \geqslant 3$) with all its vertices linked to a further point, the nub, is called a wheel. The number of spanning trees of a wheel is $L_{2n}-2$, which equals $5F_n^2$ for even n, and $5F_n^2-4$ for odd n. (See formula (23).) (See Bowcamp, 1965; Sedlaček, 1969; Myers, 1971).

12. Fibonacci numbers have also appeared in botanical studies. From the extensive literature we quote Thompson (1963, p. 922)

> In many cones we can trace five rows of scales winding steeply on the cone in one direction, and three rows winding less steeply the other way; in certain other species . . . the normal number is eight rows in one direction and five in the other . . . in some cases we shall be able to trace thirteen rows in one direction and twenty-one in the other, or perhaps twenty-one and thirty-four.

(See also Hounslow, 1973). (It will be observed that 3, 5, 8, 13, 21, 34 are successive Fibonacci numbers.)

13. Fibonacci trees are mentioned in Turner (1986). They are constructed as follows:

T_1 is a single point, and T_2 is a single point as well.

T_{n+2} is the tree obtained by putting a new node at the end of a fork starting from the latest nodes of T_n and of T_{n+1}, thus:

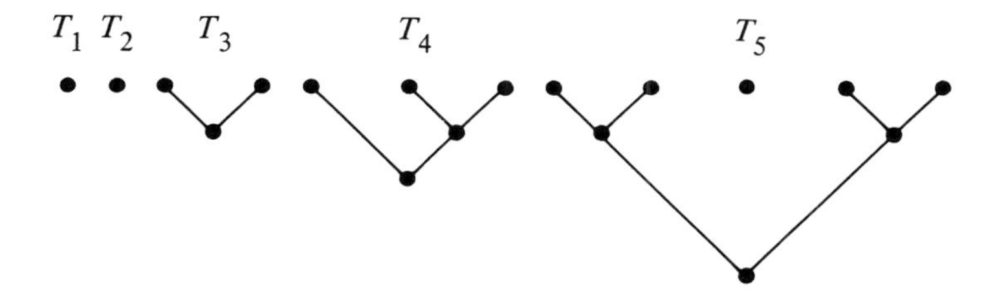

The number of nodes of the tree T_k, denoted by N_k, equals $2F_k - 1$.

Proof. We have, from the diagram,

$$N_{k+2} = N_{k+1} + N_k - 1 \qquad (k = 1, 2, 3, \ldots)$$

Start with

$$N_1 = 2F_1 - 1 \ , \qquad N_2 = 2F_2 - 1$$

(both N_1 and N_2 equal 1, by definition).

Assume that

$$N_t = 2F_t - 1 \qquad \text{and} \qquad N_{t+1} = 2F_{t+1} - 1$$

hold for $1, 2, \ldots, k$. Then

$$N_{k+2} = (2F_{k+1} - 1) + (2F_k - 1) - 1 = 2F_{k+2} - 1 \ .$$

QED.

14. Parman and Singh (1985) have pointed out that the 'so-called' Fibonacci numbers were used in India some time between the sixth and the eighth centuries A.D. to describe metres in Sanskrit and in Prakrit poetry.

II

Background

1. As suggested by the historical development, we have started with a consideration of the Fibinacci sequence. Now we set it into a more general framework.

Consider sequences of kth order, generated by the mechanism

$$u_{n+k} = a_1 u_{n+k-1} + a_2 u_{n+k-2} + \ldots + a_{k-1} u_{n+1} + a_k u_n \qquad \text{(A)}$$

for $n \geqslant 0$.

We start with a k-vector, the 'seed'

$$s_0 = (u_0, u_1, \ldots, u_{k-1})$$

and imagine it to be transformed by (A) into the vector

$$s_1 = (u_1, u_2, \ldots, u_k).$$

Generally, the mechanism (A) transforms the k-vector

$$s_n = (u_n, u_{n+1}, \ldots, u_{n+k-1})$$

into the vector

$$s_{n+1} = (u_{n+1}, u_{n+2}, \ldots, u_{n+k}).$$

We can write this transformation as

$$C.s_n = s_{n+1}$$

where C is the 'companion matrix'

$$C = \begin{pmatrix} 0 & 1 & 0 & \dots & 0 & 0 \\ 0 & 0 & 1 & \dots & 0 & 0 \\ \vdots \\ 0 & 0 & 0 & \dots & 0 & 1 \\ a_k & a_{k-1} & a_{k-2} & \dots & a_2 & a_1 \end{pmatrix}$$

Hence, in general,

$$s_t = C^t . s_0 \ .$$

If s_0 is known, then s_n can be found explicitly as follows:
Solve the characteristic equation of the difference equation (A)

$$x^k = a_1 x^{k-1} + a_2 x^{k-2} + \dots + a_k \ .$$

Let its roots be

$$r_1, r_2, \dots r_t \ .$$

If $t = k$, that is if all the roots are distinct, then

$$s_n = c_1 r_1^n + c_2 r_2^n + \dots + c_k r_k^n \ , \qquad \text{(B)}$$

where the coefficients $c_1, \dots, c_k$ depend on the seed, thus:

$$c_1 + c_2 + \dots + c_k = u_0$$
$$c_1 r_1 + c_2 r_2 + \dots + c_k r_k = u_1$$
$$\vdots$$
$$c_1 r_1^{k-1} + c_2 r_2^{k-1} + \dots + c_k r_k^{k-1} = u_{k-1} \ .$$

If r_s is not a single, but an m-fold root, then $c_s r_s^i$ in equation (B) is to be replaced by

$$c_{s_o} r_s^i + c_{s_1} i r_s^i + \dots + c_{s_{m-1}} i^{m-1} r_s^i \ . \ (i = 1, 2, \dots, k)$$

2. We consider in some more detail the case $k = 2$.
The companion matrix is now

$$\begin{pmatrix} 0 & 1 \\ a_2 & a_1 \end{pmatrix} .$$

The Fibonacci mechanism is a special case, with $a_1 = a_2 = 1$.

The roots of

$$x^2 - a_1 x - a_2 = 0$$

are $\frac{1}{2}(a_1 + \sqrt{(a_1^2 + 4a_2)}$ and $\frac{1}{2}(a_1 - \sqrt{(a_1^2 + 4a_2)}$.
We shall, in particular, be interested in the two cases

$$r_1 \neq r_2, c_1 = -c_2 = 1/(r_1 - r_2)$$

and

$$r_1 \neq r_2, c_1 = c_2 = 1 ,$$

which determine the seeds (0,1) and $(2,a_1)$ respectively.

Denote

$$(r_1^n - r_2^n)/(r_1 - r_2) \text{ by } U_n;\ U_0 = 0,\ U_1 = 1$$

and

$$r_1^n + r_2^n \text{ by } V_n;\ V_0 = 2,\ V_1 = a_1 .$$

Observe that

$$U_{2n} = U_n V_n .$$

Let us distinguish the following cases:

(i) r_1 and r_2 are distinct integers.

Example: $a_1 = 1, a_2 = 2; r_1 = 2, r_2 = -1$.

The seed (U_0,U_1) produces 0,1,1,3,5,11,21,... and the seed (V_o,V_1) produces 2,1,5,7,17,31,...

(ii) $r_1$and r_2 are conjugate complex numbers.

Example: $a_1 = 1, a_2 = -1; r_1 = \frac{1}{2}(1 + \sqrt{-3}), r_2 = \frac{1}{2}(1 - \sqrt{-3})$.

The seed (U_0,U_1) produces $0,1,1,0,-1,-1$ and the seed (V_0,V_1) produces $2,1,-1,-2,-1,1,\ldots$

These sequences are periodic, because $r_1^6 = r_2^6 = 1$.

(iii) r_1 and r_2 are real, and irrational.

Example: $a_1 = 1, a_2 = 1; r_1 = \frac{1}{2}(1+\sqrt{5}), r_2 = \frac{1}{2}(1-\sqrt{5})$.
This is the classical Fibonacci case.

The seed (U_0,U_1) produces $0,1,1,2,3,5,8,\ldots$ (Fibonacci) and the seed (V_0,V_1) produces $2,1,3,4,7,11,\ldots$ (Lucas).

Another example: $a_1 = 4, a_2 = -1; r_1 = 2+\sqrt{3}, r_2 = 2-\sqrt{3}$.

The seed (U_0,U_1) produces $0,1,4,15,56,209,\ldots$ and the seed (V_0,V_1) produces $2,4,14,52,194,\ldots$

Consider the subsequence $W_n = V_{2^n}$, that is

$$4,\ 14,\ 194,\ 37634,\ \ldots$$

The values of W_n are easily computed; we have

$$\begin{aligned}(V_{2^{n+1}}) &= [(2+\sqrt{3})^{2^{n+1}} + (2-\sqrt{3})^{2^{n+1}}] \\ &= [(2+\sqrt{3})^{2^n} + (2-\sqrt{3})^{2^n}]^2 - 2\end{aligned}$$

that is

$$W_{n+1} = W_n^2 - 2$$

The sequence is relevant in the study of Mersenne numbers $2^p - 1$, where p is an odd prime. The Mersenne number $2^p - 1$ is a prime if and only if W_{p-2} is divisible by $2^p - 1$ (Lucas 1878, p. 316).

Examples:
$p = 3$, $2^p - 1 = 7$ is a prime, since 14 is divisible by 7.
$p = 5$, $2^p - 1 = 31$ is a prime, since $37634 = 1214 \times 31$.

N.B. Of course, $2^n - 1$ can not be prime unless n is a prime, because if $n = st$, say, then $2^n - 1$ is divisible by $2^s - 1$.

When r is a double root, that is when $a_1^2 + 4a_2 = 0$, then $r = \frac{1}{2}a_1$. In such a case $u_n = (c_{10} + nc_{11})\,(\frac{1}{2}a_1)^n$.
Example:

$$a_1 = 2,\ a_2 = -1;\ r = 1.$$

$$u_n = c_{10} + nc_{11}, \text{ hence } u_o = c_{10}, u_1 = c_{10} + c_{11}.$$

When the seed is (0,1) then $c_{10} = 0$, $c_{11} = 1$, and the sequence is

0,1,2,3,4,. . .

when the seed is (2,1), then $c_{10} = 2$, $c_{11} = -1$, and the sequence is

2,1,0, − 1, − 2, − 3,. . .

3. Within this general framework, we mention now two special cases.

(a) $a_1 = a_2 = \ldots = a_k = 1$ (Miles, 1960).

The characteristic equation

$$x^k - x^{k-1} - x^{k-2} - \ldots - x - 1 = 0$$

has k distinct roots, and all of them, except one, say r_k, have absolute values less then 1. r_k lies strictly between 1 and 2.

It follows that

$$\lim_{n=\infty} \frac{u_{n+1}}{u_n} = \lim_{n=\infty} \frac{c_1 r_1^{n+1} + \ldots + c_k r_k^{n+1}}{c_1 r_1 + \ldots + c_k r_k} = r_k \ .$$

Miles (1960) contains also a generalization of our formula (54).

The sequence for $k = 3$, with

$$u_{n+3} = u_{n+2} + u_{n+1} + u_n$$

has been called the Tribonacci sequence. With seed (0,0,1) we obtain

0,0,1,1,2,4,7,13,. . .

The sequence for $k = 2$ is the Fibonacci sequence.

(b) $a_1 = a_2 = \ldots = a_{k-2} = 0, a_{k-1} = a_k = 1$ (Conolly, 1981).

With seeds (0,0,. . .,0,1) we obtain the following sequences:

$k = 3$: 0,0,1,0,1,1,1,2,2,3,4,5,7,9,12

$k = 4$: 0,0,0,1,0,0,1,1,0,1,2,1,1,3,3,2,. . .

$k = 5$: 0,0,0,0,1,0,0,0,1,1,0,0,1,2,1,0,1,3,3,1.

4. Consider the determinants

$$|A_i| = \begin{vmatrix} u_i & u_{i+1} & \cdots & u_{i+k-1} \\ u_{i+1} & u_{i+2} & \cdots & u_{i+k} \\ u_{i+2} & u_{i+3} & \cdots & u_{i+k+1} \\ \vdots & & & \\ u_{i+k-1} & u_{i+k} & & u_{i+2k-2} \end{vmatrix}$$

and

$$|A_{i-1}| = \begin{vmatrix} u_{i-1} & u_i & \cdots & u_{i+k-2} \\ u_i & u_{i+1} & \cdots & u_{i+k-1} \\ u_{i+1} & u_{i+2} & \cdots & u_{i+k} \\ \vdots & & & \\ u_{i+k-2} & u_{i+k-1} & \cdots & u_{i+2k-3} \end{vmatrix}$$

Replace in A_i each element of the first column by its row sum. This does not change the value of the determinant, so that

$$|A_i| = \begin{vmatrix} u_{i+k-1} & u_i & \cdots & u_{i+k-2} \\ u_{i+k} & u_{i+1} & \cdots & u_{i+k-1} \\ \vdots & & & \\ u_{i+2k-2} & u_{i+2k-2} & \cdots & u_{i+2k-3} \end{vmatrix} = (-1)^{k-1}|A_{i-1}| \ .$$

Hence $|A_i| = (-1)^{i(k-1)}|A_0|$ (Miles, 1960).

In our case (b) in section 3, the determinant $|A_0|$ equals

$$\begin{vmatrix} 0 & 0 & \cdots & 0 & 1 \\ 0 & 0 & \cdots & 1 & * \\ \vdots & & & & \\ 1 & * & \cdots & * & * \end{vmatrix}$$

where the entries marked * are irrelevant to the value of the determinant. In either case, for $k = 2$, this reduces to

$$|A_n| = \begin{vmatrix} F_n & F_{n+1} \\ F_{n+1} & F_{n+2} \end{vmatrix} = (-1)^{n+1} ,$$

equivalent to our formula (29) in Chapter III.

III

Relationships

In this chapter we present relationships between Fibonacci numbers. We do not aim at listing all of them — their number is well-nigh inexhaustible. We shall develop some typical groups, and mention their consequences.

1. We start with connections between Fibonacci and Lucas numbers.

$$L_{n-1} + L_{n+1} = 5F_n. \tag{5}$$

This holds for $n = 1$, and for $n = 2$, and therefore, after addition, also for $n = 3, n = 4$, and so on.

$$F_{n-1} + F_{n+1} = L_n. \tag{6}$$

Again, this holds for $n = 1$ and for $n = 2$, and hence for any integer n.

As a consequence of (5) and (6) any formula containing Fibonacci and Lucas numbers can be expressed in terms of just one of these types; however, this is not always convenient.

Using (3), (6) can also be written

$$F_{n+2} - F_{n-2} = L_n \tag{7a}$$

or also

$$F_n + L_n = 2F_{n+1}. \tag{7b}$$

2. Our next group of formulae is based on

$$G_{n+m} = F_{m-1}G_n + F_mG_{n+1} \tag{8}$$

where $G_1, G_2, \ldots$ is any generalized Fibonacci sequence.

Proof.

$$G_n = F_{-1}G_n + F_0G_{n+1} \text{ (because } F_{-1}=1,\ F_0=0).$$
$$G_{n+1} = F_0G_n + F_1G_{n+1} \text{ (because } F_0=0,\ F_1=1).$$

If (8) holds for $m=t$, and for $m=t+1$ (as it does for $t=0$), then by addition it holds for $t=2$, $t=3, \ldots$ and so on.

For $m=2$, (8) reduces to (3).

Also

$$G_{n-m} = F_{-m-1}G_m + F_{-m}G_{n+1} = (-1)^m(F_{m+1}G_n - F_mG_{n+1}). \tag{9}$$

From (8) and (9) we obtain

$$G_{n+m} + (-1)^m G_{n-m} = (F_{m-1} + F_{m+1})G_n = L_mG_n \tag{10a}$$

and

$$G_{n+m} - (-1)^m G_{n-m} = 2F_mG_{n+1} + (F_{m-1} - F_{m+1})G_n = F_m(G_{n-1} + G_{n+1}). \tag{10b}$$

We specialize now to $G_i = F_i$.

From (8) we obtain, when $m=n+1$

$$F_{n+1}^2 + F_n^2 = F_{2n+1} \tag{11}$$

while

$$F_{n+1}^2 - F_n^2 = (F_{n+1} + F_n)(F_{n+1} - F_n) = F_{n+2}F_{n-1} \tag{12}$$

and when $m=n$

$$F_{2n} = F_{n-1}F_n + F_nF_{n+1} = F_nL_n \text{ using (6).} \tag{13}$$

Therefore

$$F_{n+1}L_{n+1} - F_nL_n = F_{2n+2} - F_{2n} = F_{2n+1}. \tag{14}$$

From (11) we derive another formula, not of particular interest in itself, but useful later, when we deal with generating functions. (11) gives

$$\begin{aligned}&F_n^2 - 3F_{n-1}^2 + F_{n-2}^2 + F_{n+1}^2 - 3F_n^2 + F_{n-1}^2\\ &= F_{2n+1} - 3F_{2(n-1)+1} + F_{2(n-2)+1}\\ &= F_{2n} + F_{2n-1}) - 3F_{2n-1} + (F_{2n-1} - F_{2n-2}) = 0.\end{aligned}$$

Hence

$$(F_n^2 - 3F_{n-1}^2 + F_{n-2}^2) = -(F^2{}_{n+1} - 3F_n^2 + F_{n-1}^2)$$

and since

$$F_3^2 - 3F_2^2 + F_1^2 = 2,$$

it follows that

$$F_n^2 - 3F_{n-1}^2 + F_{n-2}^2 = 2(-1)^{n+1}. \tag{11a}$$

From (10a) we obtain

$$F_{n+m} + (-1)^m F_{n-m} = L_m F_n \tag{15a}$$

and from (10b)

$$F_{n+m} - (-1)^m F_{n-m} = F_m L_n \tag{15b}$$

Adding (15a) and (15b) produces

$$L_m F_n + L_n F_m = 2F_{n+m} \tag{16a}$$

and subtracting (15b) from (15a) gives

$$F_n L_m - L_n F_m = (-1)^m 2F_{n-m} \tag{16b}$$

If in (10a) and in (10b) we set $G_i = L_i$, then

$$L_{n+m} + (-1)^n L_{n-m} = L_m L_n \tag{17a}$$

and, using (5),

$$L_{n+m} - (-1)^m L_{n-m} = 5F_m F_n. \tag{17b}$$

If, in (17a), we set $m = n$, we obtain

$$L_{2n} + (-1)^n.2 = L_n^2. \tag{17c}$$

(17c) defines the Lucas sequence, in the sense that it does not hold if we replace L_i by any other generalized Fibonaci sequence of integers, or indeed by the ordinary Fibonacci sequence F_i. We see this as follows:

Let $G_i = x$, then (17c) means $G_2 = x^2 + 2$, hence $G_3 = x^2 + x + 2$, and $G_4 = 2x^2 + x + 4$. But from (17c) we have also $G_4 = G_2^2 - 2$, that is

$$2x^2 + x + 4 = (x^2 + 2)^2 - 2,$$

or $x^4 + 2x^2 - x - 2 = 0$.

This equation has only one integer root, viz. $x = 1$, which leads to $G_1 = 1$, $G_2 = 3$, and so on, the Lucas sequence.

3. We introduce now a general formula, from which many more relationships emerge by specialization.

Let both G_i and H_i be generalized Fibonacci sequences, and consider the expressions

$$G_{n+h}H_{n+k} - G_nH_{n+h+k} = J_n, \text{ say}$$

and

$$G_{n+1+h}H_{n+1+k} - G_{n+1}H_{n+1+h+k} = J_{n+1}.$$

We use (8) to write

$$G_{n+h} = F_{h-1}G_n + F_hG_{n+1}$$

and

$$H_{n+h+k} = F_{h-1}G_{n+k} + F_hG_{n+k+1}.$$

Then

$$\begin{aligned} J_n &= (F_{h-1}G_n + F_hG_{n+1})H_{n+k} - G_n(F_{h-1}H_{n+k} + F_hH_{n+k+1}) \\ &= F_h(G_{n+1}H_{n+k} - G_nH_{n+k+1}). \end{aligned}$$

Similarly, we use (8) to write

$$G_{n+1+h} = F_hG_n + F_{h+1}G_{n+1}$$

and

$$H_{n+1+h+k} = F_h H_{n+k} + F_{h+1} H_{n+1+k}.$$

Then

$$J_{n+1} = (F_h G_n + F_{h+1} G_{n+1})\, H_{n+k+1} - G_{n+1}(F_h H_{n+k} + F_{h+1} H_{n+1+k})$$
$$= F_h (G_n H_{n+1+k} - G_{n+1} H_{n+k}).$$

Observe that $J_{n+1} = -J_n$. Hence

$$J_1 = -J_0,\ J_2 = (-1)^2 J_0, \ldots,\ J_n = (-1)^n J_0.$$

that is

$$G_{n+h} H_{n+k} - G_n H_{n+h+k} = (-1)^n (G_n H_k - G_0 H_{n+k}). \tag{18}$$

We look now at special forms of (18).
When $G_i = L_i$ and $H_i = F_i$, then, using (16a), we obtain

$$L_{n+h} F_{n+k} - L_n F_{n+h+k} = (-1)^n (L_h F_k - 2F_{h+k}) = (-1)^{n+1} F_h L_k. \tag{19a}$$

When $G_i \equiv F_i$ and $H_i \equiv L_i$, then

$$F_{n+h} L_{n+k} - F_n L_{n+h+k} = (-1)^n F_h L_k. \tag{19b}$$

Now let $G_i \equiv H_i$. This produces

$$G_{n+h} G_{n+k} - G_n G_{n+h+k} = (-1)^n (G_h G_k - G_0 G_{h+k}).$$

In particular, when $G_i \equiv H_i \equiv F_i$, then

$$F_{n+h} F_{n+k} - F_n F_{n+h+k} = (-1)^n F_h F_k. \tag{20a}$$

This formula was mentioned, as a problem to be proved, in Danese (1960, p. 8). Its solution, found in a way different from that above, was given on p. 694 of the same volume.

For $h = -k$, an equivalent formula appears in Catalan (1886). On the other hand, when $G_i \equiv H_i \equiv L_i$, then

$$L_{n+h} L_{n+k} - L_n L_{n+h+k} = (-1)^n (L_h L_k - 2L_{h+k}).$$

If we add (17a) to (17b) we find that $2L_{h+k} = L_h L_k + 5F_h F_k$, so that

$$L_{n+h}L_{n+k} - L_nL_{n+h+k} = (-1)^{n+1}5F_hF_k. \tag{20b}$$

4. We derive further relationships by specializing subscripts. For instance, set $k=0$ in (18) and idenfiy H_i with F_i, then

$$G_{n+h}F_n - G_nF_{n+h} = (-1)^{n+1}G_0F_h.$$

Replace $n+h$ by m and use (2). We obtain

$$G_mF_n - G_nF_m = (-1)^{n+1}G_0F_{m-n} = (-1)^mG_0F_{n-m}. \tag{21}$$

d'Ocagne (1885/86) has formulated (21) as follows: (in our present notation)

$$G_m = \frac{F_mG_n + (-1)^mG_0F_{n-m}}{F_n}.$$

This formula shows that any generalized Fibonacci sequence is defined by two elements, because the other elements can be interpolated. Of course, the interpolated elements will not always turn out to be integers.

Now set $n=0$, $h=k$. We derive from (20b)

$$L_h^2 - 2L_{2h} = -5F_h^2. \tag{22}$$

It follows from (22) and (17c) that

$$L_{2h} - 2(-1)^h = 5F_h^2 \tag{23}$$

and from (23) and (17c) that

$$5F_h^2 - L_h^2 = 4(-1)^{h+1}. \tag{24}$$

5. We concentrate now on formula (24), a special case of the so-called Pell's equation. (Historians tell us that this is a misnomer, and the study of this type of Diophantine equation should more correctly be attributed to Fermat. However, we are here not concerned with this argument.)

(24) shows that F_h and L_h cannot have any common divisor larger than 2, and that either both F_h and L_h are even, or both are odd.

Write (24) for two successive values of h, and add. Then

$$5(F_h^2 + F_{h+1}^2) = L_h^2 + L_{h+1}^2 = 5F_{2h+1} \tag{25}$$

using (11).

Now consider (24) separately for odd h, and for even h, and study the solutions in integers. let $F_h = x$ and $L_h = y$.

(**a**) h is even.

$$5x^2 - y^2 = -4.$$

$(x,y) = (1,3)$ is a solution.

If (x,y) is a solution, then so is

$$(\tfrac{1}{2}(3x+y),\ \tfrac{1}{2}(5x+3y)).$$

Now,

$$\tfrac{1}{2}(3F_i + L_i) = \tfrac{1}{2}(3F_i + F_{i+1} + F_{i-1}) = \tfrac{1}{2}(3F_i + F_{i+1} + F_{i+1} - F_i) = F_{i+2} \quad (26)$$

and

$$\tfrac{1}{2}(5F_i + 3L_i) = \tfrac{1}{2}(L_{i+1} + L_{i-1} + 3L_i) = \tfrac{1}{2}(L_{i+2} - L_i + L_{i+1} - L_i + 3L_i) = L_{i+2}. \quad (27)$$

Starting with $(F_2, L_2) = (1, 3)$ we generate in this manner

(F_4L_4), (F_6,L_6), and so on, for even subscripts.

(**b**) h is odd.

$$5x^2 - y^2 = 4.$$

$(x,y) = (1,1)$ is a solution, and in precisely the same way as for even h, we generate, starting with $(F_1,L_1) = (1,1)$

(F_3,F_3), (F_5,L_5), and so on, for odd subscripts.

Altogether, then, the sets of solutions of the equations

$$5x^2 - y^2 = 4(-1)^n$$

include the pairs $(x = F_n, y = L_n)$. (Figure I.)

In fact, we can say more. The solutions which we have exhibited are the only positive integer pairs of solutions of the two Pell's equations.

Proof. The following holds for all $x \leqslant 5$: if there exists a y such that $5x^2 - y^2$ equals 4,

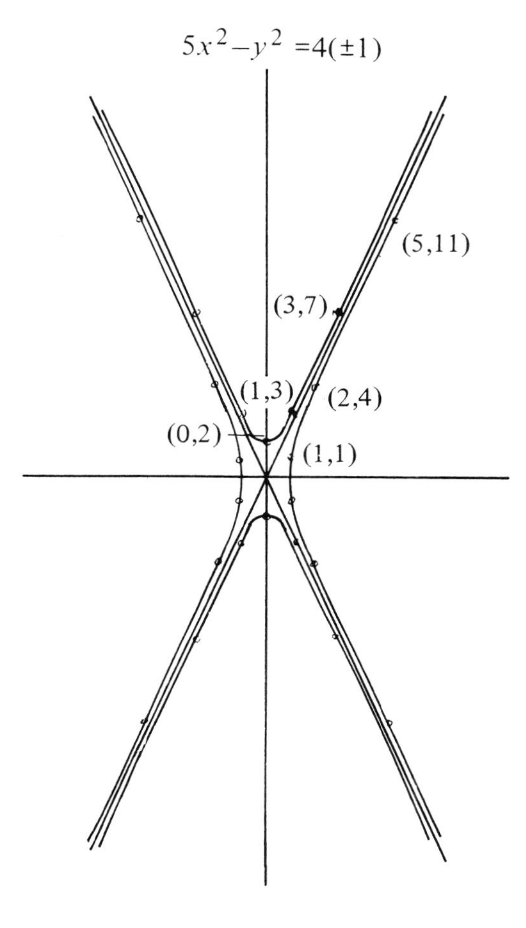

Fig. I.

or -4, then there exists a subscript n such that $x = F_n$ and $y = L_n$. (This is the case for $x = 1,2,3$ or 5; the corresponding values of y are 1,4,7 or 11.)

Now if $5x^2 - y^2 = 4$, or $= -4$, and x is larger than 1, then y is larger than x. Compute

$$x_1 = \tfrac{1}{2}(3x - y) \text{ and } y_1 = \tfrac{1}{2}(-5x + y)$$
$$x_{i+1} = \tfrac{1}{2}(3x_i - y_i) \text{ and } y_{i+1} = \tfrac{1}{2}(-5x_i + y_i).$$

This is the inverse procedure to that which we have used to generate the pairs (F_i, L_i), starting with (F_1, L_1), or with (F_2, L_2). We have $x_1 = x$, and if we repeat the computations, then we must, at some stage, reach some $x_i \leqslant 5$, and therefore $x_i = F_n$, $y_i = L_n$, for some n.

Reverse the procedure and compute from (x_i, y_i) the pairs

$$(x_{i-1},y_{i-1}),\ (x_{i-2},y_{i-2}),\ldots,(x,y).$$

At these steps we have

$$(F_{n+2},L_{n+2})\ (F_{n+1},L_{n+1}) \text{ and finally } (x,y)$$

which is again some pair (F_i,L_i).

6. Consider the matrix

$$\begin{pmatrix} G_{n+1} & G_n \\ G_n & G_{n-1} \end{pmatrix} = \begin{pmatrix} G_n & G_{n-1} \\ G_{n-1} & G_{n-2} \end{pmatrix} \begin{pmatrix} F_2 & F_1 \\ F_1 & F_0 \end{pmatrix} = \ldots = \begin{pmatrix} G_2 & G_1 \\ G_1 & G_0 \end{pmatrix} \begin{pmatrix} F_2 & F_1 \\ F_1 & F_0 \end{pmatrix}^{n-1}.$$

Replacing the matrices by their determinants, we obtain

$$G_{n+1}G_{n-1} - G_n^2 = (-1)^n(G_1^2 - G_0G_2) \tag{28}$$

and, in particular,

$$F_{n+1}F_{n-1} - F_n^2 = (-1)^n. \tag{29}$$

This our formula (20a), for $h=1$, $k=-1$.

(29) is the source of a number of significant results, and we shall now present some of them.

(a) The following curiosity seems to be of old origin, mentioned in many collections of puzzles and paradoxes. See, for instance, Ball and Coxeter (1947, pp. 84–86).

Cut a square of sides F_n into portions as in Fig. II(a) and reassemble them into a rectangle of sides F_{n-1} and F_{n+1}. The areas of the square and of the rectangle differ by 1, whatever the side of the original square (Fig. II(b)). We must therefore accept a gap or an overlap in the rectangle, dependent on whether the subscript of F_n is even or odd. Fig. II concerns the cases $n=4$ and $n=5$.

Because the gap, or the overlap, has always an area of 1, independently of n, it will be less conspicuous the larger n is. The paradox is usually formulated for $F_6=8$, where the gap is hardly noticeable. The diagram appears then to prove that $F_5F_7=63$ equals $F_6^2=64$.

If we use Lucas numbers instead of Fibonacci numbers, then the area of the gap or of the overlap will be 5 (see (20b) for $h=1$, $k=-1$.)

Generally, if we use G_i, then the gap or the overlap will be $(G_1^2-G_0G_2)$ (see formula (28)).

Similar paradoxes dependent on Fibonacci relationships are mentioned on pp. 134–137 of Gardner (1956).

(b) When n is odd, then (29) can be written

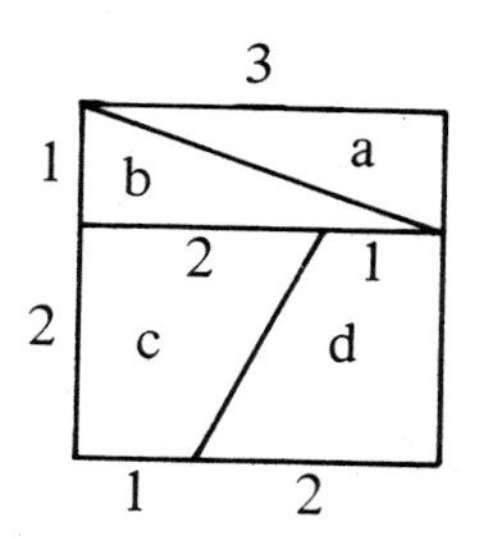

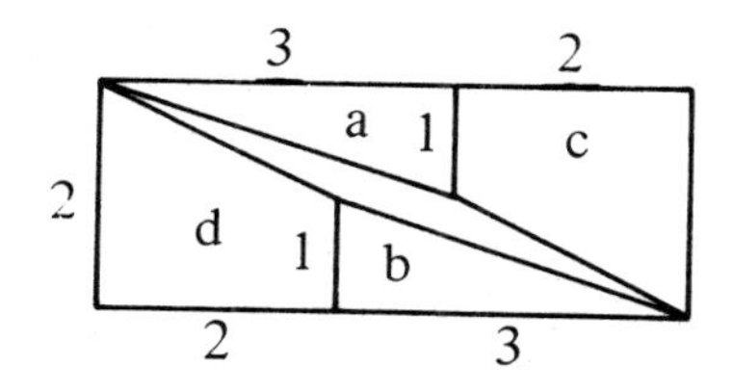

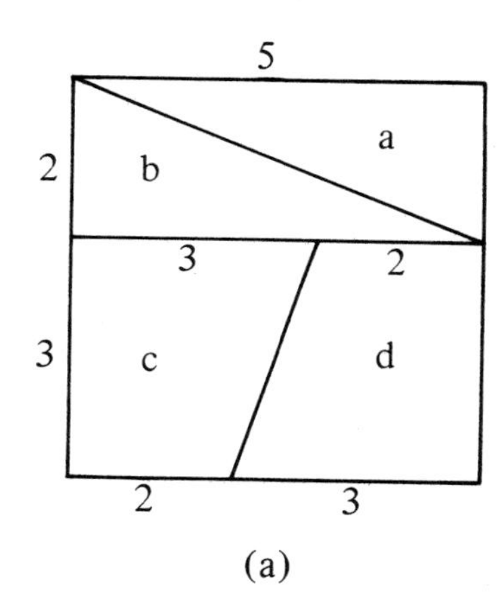

(a)

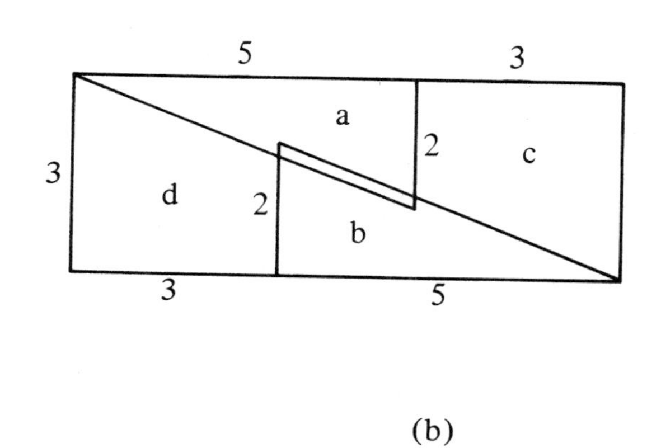

(b)

Fig. II.

$$F_{n+1}F_{n-1}+1=F_n^2.$$

Multiply both sides by F_{n+2} and then subtract F_n from both sides, remembering $F_{n+2}-F_n=F_{n+1}$. This gives

$$F_{n+1}F_{n-1}F_{n+2}+F_{n+1}=F_n^2F_{n+2}-F_n.$$

Multiply again by F_{n+2} to obtain

$$F_{n+1}F_{n+2}(F_{n-1}F_{n+2}+1)=F_nF_{n+2}(F_nF_{n+2}-1).$$

Let $F_{n+1}F_{n+2}-1=p$, $F_{n-1}F_{n+2}-1=q$, hence $p-q=\mathrm{F_nF_{n+2}}$. Then the equation above reads

$$(p+1)(q+2)=(p-q)(p-q-1).$$

Singmaster (1975) writes this in the equivalent form, using binomial coefficients

$$\binom{p+1}{q+1} = \binom{p}{q+2}$$

and this equation is solved by p and q as given above, in terms of Fibonacci numbers when n is odd. For instance, when $n=3$, then $p=14$, $q=4$, and

$$\binom{15}{5} = \binom{14}{6} = 3003.$$

(For $n=1$ the equation is trivial.)

(**c**) Write (29) as follows:

$$F_{n+1}^2 - F_{n+1}F_n - F_n^2 = (-1)^n.$$

Thus the points $(x,y) = (F_n,F_{n+1})$ lie on the two hyperbolae

$$y^2 - yx - x^2 = \pm 1$$

in the Euclidean plane. This equation can be written

$$(y - \tau x)(y + x/\tau) = \pm 1$$

where

$$\tau = \tfrac{1}{2}(1 + \sqrt{5}).$$

It shows that the two hyperbolae have the same asymptotes, orthogonal to one another, viz.

$$y - \tau x = 0 \text{ and } y + x/\tau = 0$$

(see Fig. III).
The angle arctan τ is approximately $58°17'$. The slope of the first-mentioned asymptote also indicates that the ratios (F_{n+1}/F_n) (i.e. y/x) converge to τ. This will again emerge in the chapter on continuous fractions.

We can say more about those points (y,x) on the hyperbolae: they are the only ones with positive integer coordinates on these curves. We show it as follows:

(i) Let $y^2 - xy - x^2 = 1$, $x > 0$, $y > 0$. Then we can find a positive integer i such that $x = F_{2i}$, $y = F_{2i+1}$.

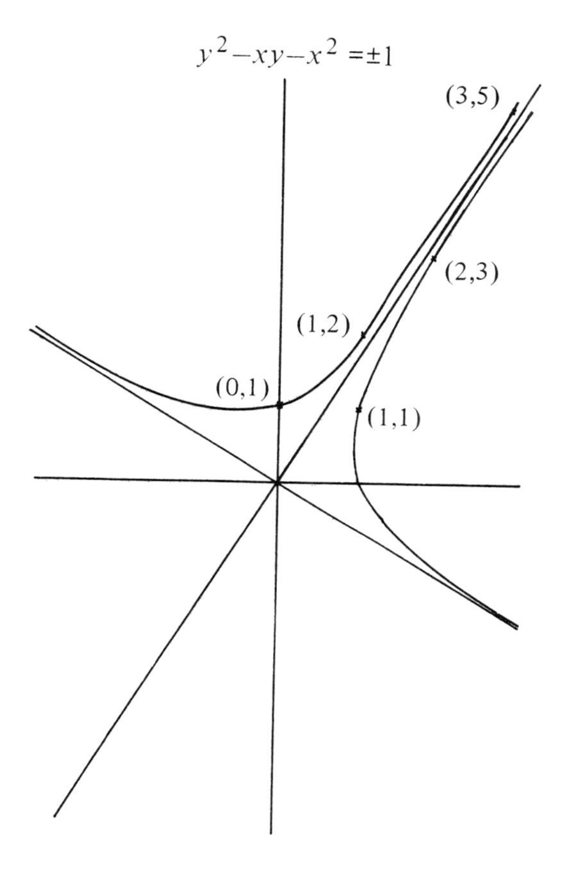

Fig. III.

Proof by induction. The statement is true for $x=1$, $y=2$ ($F_2=1$, $F_3=2$). Now let $x>1$, hence $y>2$. Assume the statement to be true for all x_0 such that $0<x_0<x$, and for the corresponding y_0. Then we can show that it is also true for x and the corresponding y.

Consider $x_0=2x-y$ and $y_0=y-x$. These values satisfy

$$y_0^2-y_0x_0-x_0^2=y^2-yx-x^2=1,$$

and we show that $0<x_0<x$ and $0<y_0$, so that the statement holds for them.

Indeed $(x+1)^2<x^2+xy+1=y^2$ (remember $y>2$) which means that $y>x+1$, $x_0=2x-y<x$, and $y_0>0$. Moreover

$$y^2=yx+x^2+1<yx+x^2+x<yx+xy+2xy$$

which means that $x_0=2x-y>0$.

We have started from the assumption that we can find i such that $x_0=F_{2i}$ and $y_0=F_{2i+1}$. But then

$$x = x_0 + y_0 = F_{2i+2} \text{ and } y = x + y_0 = F_{2i+2} + F_{2i+1} = F_{2i+3}.$$

(ii) Let $y^2 - yx - x^2 = -1, x > 0, y > 0$. Then

$$(x+y)^2 - (x+y)y - y^2 = -(y^2 - xy - x^2) = 1,$$

and we can find i such that $y = F_{2i}$, $x + y = F_{2i+1}$, hence $x = F_{2i-1}$.

Now it is but a short step to proving that the set of Fibonacci numbers is precisely the set of positive values of the polynomial

$$y(2 - [y^2 - yx - x^2]^2) = P, \text{ say},$$

for $x = 1,2,\ldots$ and $y = 1,2,\ldots$ (Jones, 1975).

For P to be positive when y is positive, we must have $y^2 - yx - x^2 = \pm 1$. (It is easy to see that $(y^2 - yx - x^2)^2$ is less than 2, and it is of course positive.) But then we can find, for any y, an x such that x and y are both Fibonacci numbers, and $y = P$. (N.B. When x and y are not a pair of consecutive Fibonacci numbers, then P is negative when x and y are both positive.)

In the same way it can be proved that the set of Lucas numbers is precisely the set of positive values of the polynomial

$$y(1 - [(y^2 - yx - x^2)^2 - 25]^2)$$

for $x = 1,2,\ldots$ and $y = 1,2,\ldots$

In this case x and y are a pair of consecutive Lucas numbers. The proof starts by writing (28) for $G_i \equiv L_i$ *as*

$$L_{n+1}^2 - L_{n+1}L_n - L_n^2 = 5(-1)^{n+1}.$$

(Jones, 1976).

(**d**) We have

$$\begin{aligned} F_{n+1}L_n &= F_{n+1}^2 + F_{n+1}F_{n-1} \text{ (by (6))} \\ &= F_{n+1}^2 + F_n^2 + (-1)^n \text{ by (29)} \\ &= F_{2n+1} + (-1)^n \text{ (by (11))} \end{aligned}$$

that is

$$F_{n+1}L_n = F_{2n+1} - 1 \text{ (for odd } n), \tag{30}$$

$$F_{n+1}L_n = F_{2n+1} + 1 \text{ (for even } n). \tag{31}$$

(e) From $F_n^2\,(F_{m+1}F_{m-1}) = F_n^2(F_m^2 + (-1)^m)$, and

$$F_m^2\,(F_{n+1}F_{n-1}) = F_m^2(F_n^2 + (-1)^n)$$

we obtain, by subtraction

$$F_n^2(F_{m+1}F_{m-1}) - F_{n+1}F_{n-1}) = (-1)^{n-1}(F_m^2 + (-1)^{m+n+1}F_n^2 = (-1)^{n-1}(F_{m+n}F_{m-n}). \qquad (32)$$

Lucas (1878) remarks that this is a formula in the theory of the theta-functions and eta-functions of Jacobi.

(f) Consider the formula

$$\arctan\,(1/F_{2m+1}) + \arctan\,(1/F_{2m+2}) = \arctan(1/F_{2m}). \qquad (*)$$

Its validity is confirmed by applying

$$\tan(\alpha+\beta) = (\tan\alpha + \tan\,\beta)/(1 - \tan\alpha\,\tan\,\beta),$$

thus

$$\tan[\arctan(1/F_{2m+1}) + \arctan(1/F_{2m+2})] = \left(\frac{1}{F_{2m+1}} + \frac{1}{F_{2m+2}}\right)\Big/\left(1 - \frac{1}{F_{2m+1}F_{2m+2}}\right) = \frac{1}{F_{2m}}$$

that is, after multiplying both sides by $F_{2m}F_{2m+1}F_{2m+2}$

$$F_{2m}F_{2m+1} + F_{2m}F_{2m+2} = F_{2m+1}F_{2m+2} - 1$$

or, since

$$F_{2m+1}F_{2m+2} - F_{2m}F_{2m+1} = F_{2m+1}^2,$$
$$F_{2m}F_{2m+2} = F_{2m+1}^2 - 1.$$

This is again (29), with $2m+1$ replacing n.

It follows from (*) that

$$\sum_{m=1}^{\infty} \arctan\,(1/F_{2m+1}) = (\arctan(1/F_2) - \arctan(1/F_4)) + (\arctan(1/F_4) - \arctan(1/F_6)) + \ldots = \arctan 1 = \tfrac{1}{4}\pi,$$

a formula which relates π and Fibonacci numbers in a simple, though at a first glance surprising, manner.

(**g**) J.C. Owings jr (1987) has proved that the pairs of integers (a,b), $(1 \leqslant a \leqslant b)$, which satisfy

$$a^2 \equiv -1 (\text{mod } b), \text{ and } b^2 \equiv -1 (\text{mod } a)$$

are the numbers (F_{2t-1}, F_{2t+1}) $(k = 1,2,\ldots)$.

This can be shown to be a consequence of (29), setting $h = 2$, $k = -2$ and

$$\text{(a) } n = 2t + 1 \text{ or (b) } n = 2t - 1.$$

We obtain then

$$\text{(a) } F_{2t+3}F_{2t-1} = F_{2t+1}^2 + 1 \text{ and (b) } F_{2t+1}F_{2t-3} = F_{2t-1}^2 + 1.$$

Hence

$$\text{(a) } F_{2t+1}^2 \equiv -1 \ (\text{mod } F_{2t-1}), \text{ and}$$
$$\text{(b) } F_{2t-1}^2 \equiv -1 \ (\text{mod } F_{2t+1}).$$

This shows that the pair (F_{2t-1}, F_{21t+1}) has the required property. It remains to show that they are the only such pairs.

Now if $(a,b) = (F_{2t-1}, F_{2t+1})$ has this property, then so has $(c,a) = (F_{2t-3}, F_{2t-1})$, where $c = (a^2 + 1)/b$, since $c^2 \equiv -(a^2 + 1)^2 \equiv -1$ (mod a), and $1 \leqslant c \leqslant a$, with $c = a$ only when $a = c = 1$.

We can thus reduce the numbers in the pair until we reach $(1,1) = (F_{-1},F_1)$.

Then we can trace our way back, from any (m,n) to $(n,(n^2 + 1)/m)$ through $(1,2)$, $2,5)$ . . . until we reach again

$$(a,b) = (F_{2t-1}, F_{2t+1}),$$

and all intermediate pairs have the same structure. This completes the proof.

7. We shall now consider sums of generalized Fibonacci numbers.

$$\sum_{i=1}^{n} G_i = \sum_{i=3}^{n+2} G_i - \sum_{i=2}^{n+1} G_i = (\sum_{i=1}^{n+2} G_i - G_1 - G_2) - (\sum_{i=1}^{n+2} G_i - G_1 - G_{n+2}),$$

hence

$$\sum_{i=1}^{n} G_i = G_{n+2} - G_2. \tag{33}$$

Also

$$\sum_{i=1}^{n} G_{2i-1} = \sum_{i=1}^{n} G_{2i} - \sum_{i=1}^{n} G_{2i-2} = \sum_{i=1}^{n} G_{2i} - \sum_{i=0}^{n-1} G_{2i},$$

hence

$$\sum_{i=1}^{n} G_{2i-1} = G_{2n} - G_0. \tag{34}$$

Similarly

$$\sum_{i=1}^{n} G_{2i} = \sum_{i=1}^{n} G_{2i+1} - \sum_{i=1}^{n} G_{2i-1} = \sum_{i=1}^{n} G_{2i+1} - \sum_{i=0}^{n-1} G_{2i+1},$$

hence

$$\sum_{i=1}^{n} G_{2i} = G_{2n+1} - G_1. \tag{35}$$

Addition of (34) and (35) reproduces (33), writing $2n$ for n. Subtraction of (34) from (35) produces

$$\sum_{i=1}^{n} G_{2i} - \sum_{i=1}^{n} G_{2i-1} = \sum_{i=1}^{2n} (-1)^i G_i = G_{2n-1} + G_0 - G_1. \tag{36}$$

We introduce now a sum, the limit of which as n tends to infinity will be of interest in the next Chapter (for $G \equiv F$).

$$\sum_{k=1}^{n} G_{k-1}/2^k = \tfrac{1}{2}(G_0 + G_3) - G_{n+2}/2^n. \tag{37}$$

This is easily proved by induction. The formula is obviously true for $n = 1$.

Assume it to be true for $2,3,\ldots,n-1$. Add $G_{n-1}/2^n$ on both sides. The right-hand side becomes

$$\tfrac{1}{2}(G_0+G_3)-(2G_{n+1}-G_{n-1})/2^n=\tfrac{1}{2}(G_0+G_3)-(G_{n+1}+G_n)/2^n$$
$$=\tfrac{1}{2}(G_0+G_3)-G_{n+2}/2^n \text{ QED.}$$

When $G_i \equiv F_i$, then (37) reduces to

$$\sum_{k=1}^{n} F_{k-1}/2^k = 1 - F_{n+2}/2^n. \tag{37a}$$

The next formula which we derive is the basis for some amusement, as we shall explain.

In (13), replace n by the odd number $2k+1$, to give

$$F_{2k+1}L_{2k+1}=F_{4k+2}.$$

In the same way, obtain from (30)

$$F_{2k+2}L_{2k+1}=F_{4k+3}-1.$$

Now replace in (33) n by $4k+2$, thus

$$\sum_{i=1}^{4k+2} G_i = G_{4k+4}-G_2.$$

Because of (8),

$$G_{4k+4}=G_{1+(4k+3)}=F_{4k+2}G_1+F_{4k+3}G_2.$$

Thus

$$\sum_{i=1}^{4k+2} G_i = F_{4k+2}G_1+(F_{4k+3}-1)G_2$$
$$= L_{2k+1}(F_{2k+1}G_1+F_{2k+2}G_2), \text{ that is}$$
$$\sum_{i=1}^{4k+2} G_i = L_{2k+1}G_{2k+3}. \tag{38}$$

For instance, for $k=1$ $\sum_{i=1}^{6} G_i = 4G_5$, for $k=2$ $\sum_{i=1}^{10} G_i = 11G_7$.

The second of these formulae has been proposed as a means of surprising an audience by the speed with which 10 successive terms of any generalized Fibonacci sequence can be added.

8. We turn now to expressions with sums of second-order terms. (28) can be written

$$\begin{aligned}&(G_n + G_{n-1})G_{n-1} - G_n^2 = (-1)^n(G_1^2 - G_0G_2)\\ &= G_nG_{n-1} + G_{n-1}^2 - G_n^2 = (-1)^n(G_1^2 - G_0G_2).\end{aligned}$$

Thus

$$\begin{aligned}&G_1G_0 + G_0^2 - G_1^2 = (-1)(G_1^2 - G_0G_2)\\ &G_2G_1 + G_1^2 - G_2^2 = (-1)^2(G_1^2 - G_0G_2)\\ &\cdot\\ &\cdot\\ &\cdot\\ &G_{2n}G_{2n-1} + G_{2n-1}^2 - G_{2n}^2 = (-1)^{2n}(G_1^2 - G_0G_2).\end{aligned}$$

By addition, (and because $\sum_{i=1}^{2n} (-1)^i = 0$)

$$\sum_{1}^{2n} G_iG_{i-1} = G_{2n}^2 - G_0^2. \tag{39}$$

For $G_i \equiv F_i$ this means

$$\sum_{i=1}^{2n} F_iF_{i-1} = F_{2n}^2. \tag{40}$$

We illustrate this formula in Fig. IV(a).
If we add the further equation

$$G_{2n+1}G_{2n} + G_{2n}^2 - G_{2n+1}^2 = (-1)^{2n+1}(G_1^2 - G_0G_2)$$

then

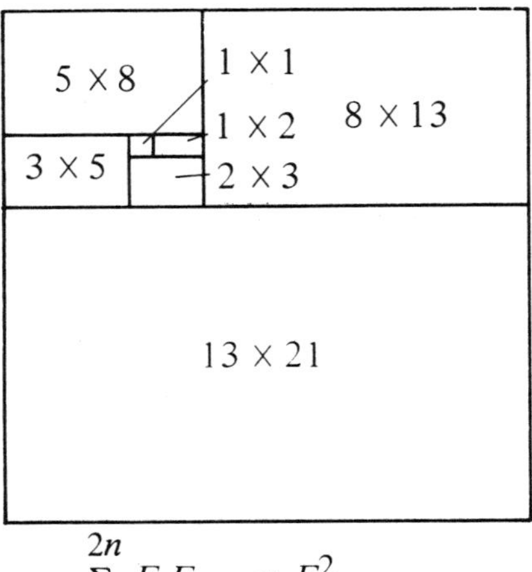

$$\sum_{i=1}^{2n} F_i F_{i-1} = F_{2n}^2$$

Fig. IV(a).

$$\sum_{i=1}^{2n+1} G_i G_{i-1} = G_{2n+1}^2 - G_0^2 - (G_1^2 - G_0 G_2) \tag{41}$$

and

$$\sum_{i=1}^{2n+1} F_i F_{i-1} = F_{2n-1}^2 - 1. \tag{42}$$

Fig. IV(b) illustrates this case.

9. Consider

$$G_{n+1}^2 - G_n^2 = (G_{n+1} - G_n)(G_{n+1} + G_n) = G_{n+2}G_{n-1}.$$

Thus

$$\begin{aligned} G_2^2 - G_1^2 &= G_3 G_0 \\ G_3^2 - G_2^2 &= G_4 G_1 \\ &\cdot \\ &\cdot \\ &\cdot \\ G_{n+1}^2 - G_n^2 &= G_{n+2} G_{n-1} \end{aligned}$$

and, after addition,

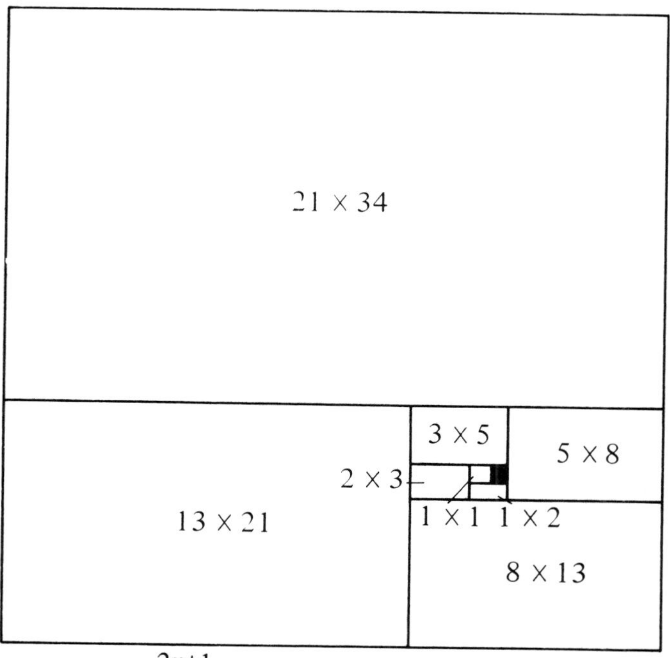

Fig. IV(b).

$$\sum_{i=1}^{n} G_{i+2}G_{i-1} = G_{n+1}^2 - G_1^2. \tag{43}$$

10. We have

$$\begin{aligned}\sum_{i=1}^{n} G_i^2 = \sum_{i=1}^{n} G_i(G_{i+1} - G_{i-1}) &= \sum_{i=1}^{n} G_iG_{i+1} - \sum_{i=1}^{n} G_{i-1}\, G_i \\ &= (G_1G_2 - G_0G_1) + (G_2G_3 - G_1G_2) + \ldots + (G_nG_{n+1} - G_{n-1}G_n),\end{aligned}$$

that is

$$\sum_{i=1}^{n} G_i^2 = G_nG_{n+1} - G_0G_1. \tag{44}$$

If $G_i \equiv F_i$, then

$$\sum_{i=1}^{n} F_1^2 = F_n F_{n+1}. \tag{45}$$

This formula can also be illustrated geometrically. We do it in Fig. V(a). (See

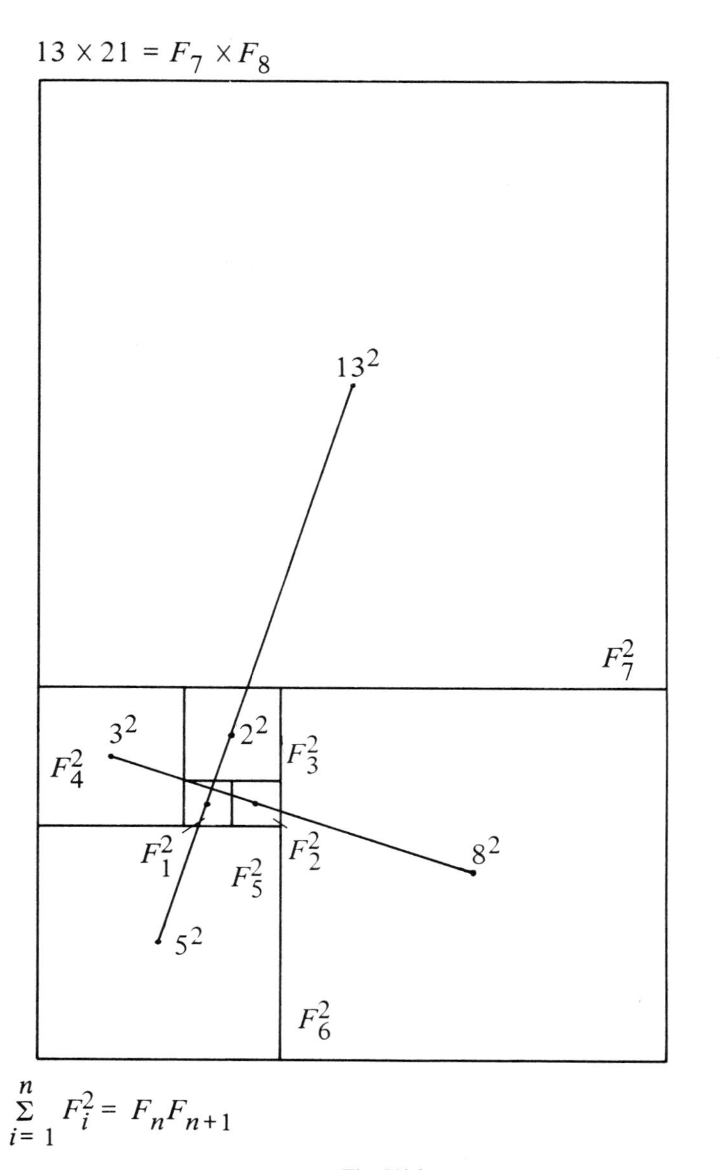

Fig. V(a).

Holden, 1975.)

As n increases, the rectangle tends to a Gold Rectangle, because $\lim_{n=\infty} \frac{F_{n+1}}{F_n}$ tends to τ (see (101) in Chapter VIII).

In Fig. V(a) we have also drawn the straight lines which connect the centres C_i of alternate squares.

Let C_1, the centre of the F_1-square be the origin of a system of coordinates with horizontal and vertical axes. Then the C_i, the centres of the F_i-squares, have coordinates as follows:

$$\begin{array}{llll} C_1\ (0,0) & C_3\ (\tfrac{1}{2},\tfrac{1}{2}) & C_5\ (-1,-3) & C_7\ (3,9) \\ C_2\ (1,0) & C_4\ (-2,1) & C_6\ (5\tfrac{1}{2},-1\tfrac{1}{2}) & C_8\ (-14,5) \end{array}$$

The points $C_1, C_3, \ldots$ lie on the straight line $y = 3x$, and the points $C_2, C_4, \ldots$ lie on the straight line $y = \frac{1-x}{3}$. This is so for the points quoted above, and must remain true, as the diagram expands. The two lines are orthogonal and intersect at (1/10, 3/10).

We can also say something about the distances of the C_i from the point of intersection.

Take any three successive squares (apart from the first), such as the F_2, F_3 and F_4 squares. Denote their centres by A, B and C respectively, and denote the point on the line AC where it meets the perpendicular from B by M (Fig. V(b)). Then a

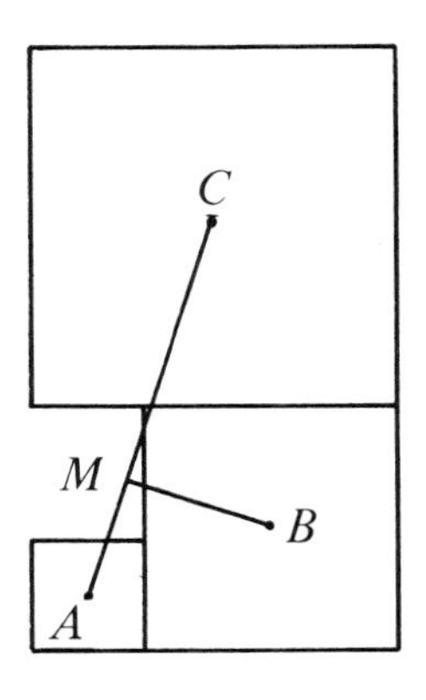

Fig. V(b).

straightforward application of coordinate geometry shows that $AM + BM = CM$. Thus the distances of the centres from the point of intersection of the two orthogonal lines form a generalized Fibonacci sequence.

To identify the sequence, we compute the distance of the intersection from C_1 and from C_2 and find that they are $1/\sqrt{10}$ and $3/\sqrt{10}$. The sequence is (the multiple of) a Lucas sequence.

In an analogous diagram (Fig. V(c)) to illustrate the formula $\sum_{i=1}^{n} L_i^2 = L_n L_{n+1} - 2$, we can again identify two orthogonal lines with slopes 3 and $-1/3$ connecting centres of squares, and the distances of these centres from the point of intersection of the two lines form (the multiple of) a Fibonacci sequence.

10. Observe that

$$G_{n+p} = \sum_{i=0}^{p} \binom{p}{i} G_{n-i} \tag{46}$$

holds for all integer n when $p = 1$. We shall prove, by induction, that it holds for every positive integer p.

Let (46) be true up to some value of p. Then

$$\begin{aligned}
G_{m+p} &= \sum_{i=0}^{p} \binom{p}{i} G_{m-i} = \sum_{i=0}^{p} \binom{p}{i} (G_{m-i-1} + G_{m-i-2}) \\
&= G_{m-1} + \sum_{i=1}^{p} \binom{p}{i} G_{m-i-1} + \sum_{i=0}^{p-1} \binom{p}{i} G_{m-i-1} + G_{m-p-2} \\
&= G_{m-1} + \sum_{i=1}^{p} \binom{p}{i} G_{m-i-1} + \sum_{i=1}^{p} \binom{p}{i-1} G_{m-i-1} + G_{m-p-2} \\
&= G_{m-1} + \sum_{i=1}^{p} \binom{p+1}{i} G_{m-i-1} + G_{m-p-2} \\
&= \sum_{i=0}^{p+1} \binom{p+1}{i} G_{m-i-1}.
\end{aligned}$$

Let $m = n + 1$, then

$$G_{n+p+1} = \sum_{i=0}^{p+1} \binom{p+1}{i} G_{n-i}.$$

This is again (46), but p replaced by $p + 1$. Thus (46) holds for all p.

Because $\binom{p}{i} = \binom{p}{p-i}$, (46) can also be written

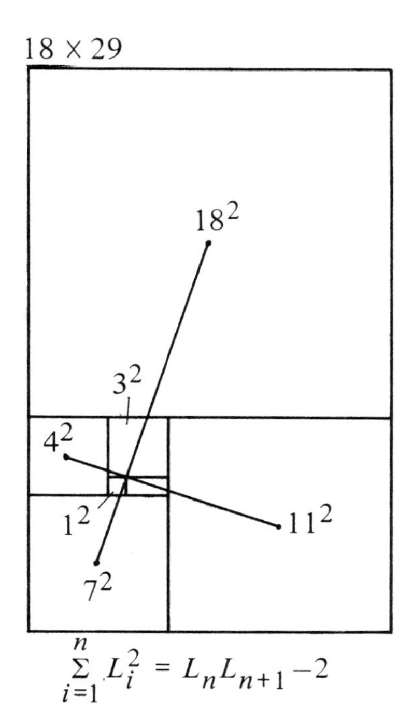

Fig. V(c).

$$G_{n+p} = \sum_{i=0}^{p} \binom{p}{i} G_{n-p+i}.$$

When $n = p$, then

$$G_{2n} = \sum_{i=0}^{n} \binom{n}{i} G_i. \tag{47}$$

When $n = m + (t-1)p$, then

$$G_{m+tp} = \sum_{i=0}^{p} \binom{p}{i} G_{m+(t-2)p+1}. \tag{48}$$

In particular,

$$G_{m+2p} = \sum_{i=0}^{p} \binom{p}{i} G_{m+i}. \tag{49}$$

For $m = 1$, $p = n$, $G_i \equiv F_i$ (49) can be transformed into

$$F_{2n+1} = 1 + \sum_{i=0}^{n} \binom{n+1}{i+1} F_i \tag{50}$$

as follows:

$$\begin{aligned} F_{2n+1} &= \sum_{i=0}^{n} \binom{n}{i} F_{i+1} = F_1 + \sum_{i=1}^{n} \binom{n}{i} (F_i + F_{i-1}) \\ &= F_1 + \sum_{i=1}^{n} \binom{n}{i} F_i + \sum_{i=1}^{n+1} \binom{n}{i+1} F_i = F_1 + \sum_{1}^{n-1} \binom{n+1}{i+1} F_i + F_n \\ &= 1 + \sum_{i=0}^{n} \binom{n+1}{i+1} F_i. \end{aligned}$$

Now observe that

$$G_{n-p} = \sum_{i=0}^{p} (-1)^i \binom{p}{i} G_{n+p-i} \tag{51}$$

holds for $p = 1$.

Write

$$\begin{aligned} G_{m-p} &= \sum_{i=0}^{p} (-1)^i \binom{p}{i} (G_{m+p+2-i} - G_{m+p+1-i}) \\ &= G_{m+p+2} + \sum_{i=1}^{p} \Bigg[(-1)^i \binom{p}{i} G_{m+p+2-i} + \\ &\quad + (-1)^i \binom{p}{i-1} G_{m+p+2-i} \Bigg] + (-1)^{p+1} G_{m+1} \\ &= \sum_{i=0}^{p+1} (-1)^i \binom{p+1}{i} G_{m+p+2-i}. \end{aligned}$$

Set $m = n - 1$, then

$$G_{n-p-1} = \sum_{i=0}^{p+1} (-1)^i \binom{p+1}{i} G_{n+p+1-i}$$

which is the same as (51), with $p+1$ replacing p. Hence (51) is proved.
(51) can be written

$$G_{n-p} = \sum_{i=0}^{p} (-1)^{p-i} \binom{p}{i} G_{n+i}.$$

Set $n = m - (t-1)p$, then

$$G_{m-tp} = \sum_{i=0}^{p} (-1)^i \binom{p}{i} G_{m-(t-1)p+i}. \tag{52}$$

In particular, for $t = 2$

$$G_{m-2p} = \sum_{i=0}^{p} (-1)^i \binom{p}{i} G_{m-p+i} = \sum_{i=0}^{p} (-1)^{p-i} \binom{p}{i} G_{m-i}. \tag{53}$$

11. We derive now

$$F_n = \sum_{i=1}^{\infty} \binom{n-i-1}{i}. \tag{54}$$

(Only a finite number of these binomial coefficients differ from zero.)
We have $F_1 = 1$ and $F_2 = 1$. Let

$$F_m = \sum_{i=0}^{\infty} \binom{m-i-1}{i} \text{ and } F_{m+1} = \sum_{i=0}^{\infty} \binom{m-i}{i} = 1 + \sum_{i=0}^{\infty} \binom{m-i-1}{i+1}.$$

Now

$$\binom{m-i-1}{i} + \binom{m-i-1}{i+1} = \binom{m-i}{i+1},$$

so that

$$F_{m+2} = F_{m+1} + F_m = 1 + \sum_{i=0}^{\infty} \binom{m-i}{i+1} = \sum_{i=0}^{\infty} \binom{m-i+1}{i}.$$

Therefore, if (54) holds for m, and for $m+1$, then it holds for $m+2$ as well. We have proved (54) by induction.

In the same way the mechanism

$$S_{n+2} = a_1 S_{n+1} + a_2 S_n$$

produces, when the seed is (0,1),

$$S_n = a_1^{n-1} + \binom{n-2}{1} a_1^{n-3} a_2 + \binom{n-3}{2} a_1^{n-5} a_2^2 \ldots (n \geqslant 1) \tag{54a}$$

(Siebeck, 1846).

IV

Fibonacci numbers and the Golden Section

1. Any number G_n can be worked out by starting from G_1 and G_2, (or from G_0 and G_1) and using (3) repeatedly. However, it is natural that we should wish to be able to find G_n without having to work out all the terms which precede it.

This can be done, of course, by using formula (54), but this would still involve the addition of $\frac{1}{2}(n+1)$ terms if n is odd, or of $\frac{1}{2}n$ terms if n is even.

A more useful method starts from the 'difference equation' (3).

Let us assume that the solution of this equation has the form x^i, where x has still to be determined. Then

$$x^{n+2} - x^{n+1} - x^n = 0.$$

We are not interested in the root $x = 0$, which would lead to the trivial Fibonacci sequence $0, 0, \ldots$, so we may divide by x^n and consider the quadratic equation

$$x^2 - x - 1 = 0.$$

This equation has two roots, $\frac{1}{2}(1+\sqrt{5}) = \tau$, say, and $\frac{1}{2}(1-\sqrt{5}) = \sigma$, say. (A number of other notations appear in the literature.) Observe that τ is positive, and that σ is negative. The number

$$1.6180339887\ldots$$

is the Golden-Section number; the reason for this name will be traced in Chapter XIII. (This is different from the Golden Number, a term use in the construction of the ecclesiastical calendar.) We note here a few facts which we shall frequently use in this book:

$$\tau + \sigma = 1, \quad \tau - \sigma = \sqrt{5}, \quad \tau\sigma = -1, \quad \tau^2 = 1 + \tau, \quad \sigma^2 = 1 + \sigma.$$

G_n may be τ^n, or σ^n, or indeed any linear combination

$$G_n = \alpha\tau^n + \beta\sigma^n. \tag{55}$$

The values α and β depend on the particular sequence we are dealing with. Since

$$G_0 = \alpha + \beta, \quad G_1 = \alpha\tau + \beta\sigma,$$

we have

$$\alpha = (G_1 - G_0\sigma)/\sqrt{5} \quad \text{and} \quad \beta = (G_0\tau - G_1)/\sqrt{5}. \tag{56}$$

We note, also, that

$$\alpha\beta = (G_0G_1 + G_0^2 - G_1^2)/5 = (G_0G_2 - G_1^2)/5. \tag{57}$$

For the Fibonacci sequence $F_0=0$, $F_1=1$, hence $\alpha=1/\sqrt{5}$, $\beta=-1/\sqrt{5}$. For the Lucas sequence, $L_0=2$, $L_1=1$, hence $\alpha=1$, $\beta=1$. Thus

$$F_n = (\tau^n - \sigma^n)/\sqrt{5} \quad \text{and} \tag{58}$$

$$L_n = \tau^n + \sigma^n. \tag{59}$$

Formula (58) was known to De Moivre (1718) and rediscovered by Binet (1843), and also by Lamé (1844) after whom the sequence is sometimes called.

Fibonacci numbers as well as Lucas numbers are integers. τ^n and σ^n are, of course, also generalized Fibonacci numbers, though integers only for $n=0$. Nevertheless, formula (8) applies to them as well. For instance, with $n=0$,

$$\tau^m = \tau F_m + F_{m-1}.$$

From (58) and (59) it follows, trivially, that

$$\tau^n = \tfrac{1}{2}(L_n + \sqrt{5}F_n) \quad \text{and} \quad \sigma^n = \tfrac{1}{2}(L_n - \sqrt{5}F_n).$$

From (56) we notice that α and β have the form

$$\alpha = \frac{m}{2} + \frac{n\sqrt{5}}{10} \quad \text{and} \quad \beta = \frac{m}{2} - \frac{n\sqrt{5}}{10}$$

where $m=G_0$ and $n=2G_1-G_0$ are integers. So $G_0=m$ and $G_1=\frac{1}{2}(m+n)$, which shows that for G_0 and G_1, and hence for all G_i to be integers, it is necessary and sufficient that m and n be either both even, or both odd integers. For Fibonacci numbers $m=0$, $n=2$, and for Lucas numbers $m=2$, $n=0$.

2. A. De Moivre established (58) by using the 'generating function'

$$g(x) = \sum_{i=0}^{\infty} F_i x^i.$$

We have

$$\begin{aligned} g(x) - F_0x^0 - F_1x^1 &= g(x) - x \\ &= \sum_{i=2}^{\infty} F_i x^i = \sum_{i=2}^{\infty} (F_{i-1}x^i + F_{i-2}x^i) \\ &= x\sum_{i=0}^{\infty} F_i x^i + x^2 \sum_{i=0}^{\infty} F_i x^i = xg(x) + x^2 g(x) \end{aligned}$$

or

$$g(x) = \frac{x}{1-x-x^2}.$$

Now

$$g(x) = \frac{x}{(1-\tau x)(1-\sigma x)} = \frac{1}{\sqrt{5}(1-\tau x)} - \frac{1}{\sqrt{5}(1-\sigma x)},$$

hence

$$g(x) = \frac{1}{\sqrt{5}}(1+\tau x+\tau^2x^2+\tau^3x^3\ldots) - \frac{1}{\sqrt{5}}(1+\sigma x+\sigma^2x^2+\sigma^3x^3+\ldots)$$

$$= [(\tau-\sigma)x+(\tau^2-\sigma^2)x^2+(\tau^3-\sigma^3)x^3+\ldots]/\sqrt{5}.$$

The coefficient of x^n, that is F_n is $(\tau^n-\sigma^n)/\sqrt{5}$, as in (58).

The corresponding formula for Lucas numbers can be found from the generating function

$$h(x) = \sum_{i=0}^{\infty} L_i x^i$$

as follows:

$$h(x)-2-x = \sum_{i=2}^{\infty} L_i x^i = \sum_{i=2}^{\infty}(L_{i-1}x^i + L_{i-2}x^i)$$

$$= x\sum_{i=1}^{\infty} L_i x^i + x^2\sum_{i=0}^{\infty} L_i x^i$$

$$= x(h(x)-2)+x^2h(x).$$

This means

$$h(x) = \frac{2-x}{1-x-x^2} = \frac{1}{1-\tau x} + \frac{1}{1-\sigma x}$$

or

$$h(x) = 2+(\tau+\sigma)x+(\tau^2+\tau^2)x^2+\ldots$$

and thus establishes (59).

Let $x=\frac{1}{2}$. Then, equating $\sum_{i=0}^{\infty} F_i x^i$ and $g(x)$ numerically, that is

$$1/(1-x-x^2) = \sum_{i=0}^{\infty} F_i x^{i-1} \qquad (*)$$

we obtain

$$\sum_{i=1}^{\infty} F_i/2^{i-1} = 4, \quad \text{or}$$

$$\sum_{i=1}^{\infty} F_i/2^i = 2. \tag{60}$$

Differentiate the two sides of (*) with regard to x.

$$\frac{1+2x}{(1-x-x^2)^2} = \sum_{i=1}^{\infty} (i-1)F_i x^{i-2}$$

and when $x = \frac{1}{2}$, $8 = \sum_{i=1}^{\infty} (iF_i/2^i) - 2$, that is

$$\sum_{i=1}^{\infty} iF_i/2^i = 10. \tag{61}$$

3. We are now in a position to supply the proofs for the two probabilistic statements in Chapter I. We repeat them here for convenience.

(i) A fair coin is tossed until two consecutive heads (HH) appear. The probability of the tosses of having k terms is $F_{k-1}/2^k$.

Proof. For $k=2$ the probability is clearly $\frac{1}{4}$, and for $k=1$ it is 0. For $k \geqslant 3$, a sequence of length k is either a sequence of length $k-1$ preceded by tail (T), or one of length $k-2$, preceded by HT. Therefore the probability of a sequence of length k, P_k say, is

$$P_k = \tfrac{1}{2}P_{k-1} + \tfrac{1}{4}P_{k-2}.$$

We solve this difference equation by considering $x^2 - \frac{1}{2}x - \frac{1}{4} = 0$, that is $x = \frac{1}{2}\tau$, or $x = \frac{1}{2}\sigma$, and $P_k = \alpha(\frac{1}{2}\tau)^k + \beta(\frac{1}{2}\sigma)^k$. Because of $P_1 = 0$, $P_2 = \frac{1}{4}$, we have

$$\alpha = 1/(\tau\sqrt{5}) \quad \text{and} \quad \beta = -1/(\sigma\sqrt{5})$$

so that

$$P_k = (\tau^{k-1} - \sigma^{k-1})/(2^k\sqrt{5}) = F_{k-1}/2^k \quad \text{QED}$$

Observe that $\sum_{k=2}^{\infty} F_{k-1}/2^k = 1$. This follows from (37), because

$$\lim_{n=\infty} F_{n+2}/2^n = 0.$$

The expected value of the length of such a sequence is

$$\sum_{k=1}^{\infty} kF_{k-1}/2^k = \sum_{k=0}^{\infty} (k+1)F_k/2^{k+1} = 6,$$

from (60) and (61).

(ii) A fair coin is tossed until either three consecutive heads (HHH) or three consecutive tails (TTT) appear. The probability of the sequence having length k is $F_{k-2}/2^{k-1}$.

This appeared as Problem 17.8 on p. 19 of *Mathematical Spectrum* **17**, 1984/85. We give here the solution from p. 60, **18** 1985/86.

Denote the probability of the sequence terminating after k terms by $P(k)$, that of its terminating after k terms, given that the first toss produced H, by $Q(k)$, and that of its terminating after k terms given that the first toss produced T by $R(k)$. Then

$$P(k) = Q(k) + R(k).$$

Clearly, $P(1) = P(2) = 0$, $P(3) = \frac{1}{4}$. Let $k \geqslant 4$.

If, to start with, we had HH, then the third toss must produce T, otherwise we would not have $k \geqslant 4$. Therefore, after 2 tosses, $k-2$ tosses remain, starting with T. On the other hand, if the first toss was H and the second was T, then T is the first of a remaining sequence of $k-1$ tosses. Hence.

$$Q(k) = \tfrac{1}{4}R(k-2) + \tfrac{1}{2}R(k-1).$$

Similarly,

$$R(k) = \tfrac{1}{4}Q(k-2) + \tfrac{1}{2}Q(k-1).$$

Adding, we obtain

$$P(k) = \tfrac{1}{4}P(k-2) + \tfrac{1}{2}P(k-1).$$

This is the difference equation we had in (i), so that again

$$x = \tfrac{1}{2}\tau, \quad \text{or} \quad x = \tfrac{1}{2}\sigma.$$

Therefore

$$P(k) = \alpha(\tfrac{1}{2}\tau)^k + \beta(\tfrac{1}{2}\sigma)^k.$$

Because of $P(2) = 0$, $P(3) = \frac{1}{4}$, we have

$$\alpha = 2/(\sqrt{5}\tau^2) \quad \text{and} \quad \beta = -2/(\sqrt{5}\sigma^2),$$

hence

$$P(k) = F_{k-2}/2^{k-1}.$$

Again,

$$\sum_{k=3}^{\infty} F_{k-2}/2^{k-1} = 1.$$

The expected value of the length of a sequence is

$$\sum_{k=0}^{\infty} kF_{k-2}/2^{k-1}$$

which equals (using (60) and (61))

$$\sum_{k=0}^{\infty} (k+2)F_k/2^{k+1} = 7.$$

Similar methods can be applied when the number 2 in question (i), or the number 3 in question (ii), is replaced by some other number, but then the answer does not involve Fibonacci numbers (cf. Forfar and Keogh, 1985/86). For instance, the solution to question (i) for three consecutive heads demands the solution of $x^3 - x^2 - x - 1 = 0$, which produces the 'Tribonacci' sequence.

We can also derive a generating function of F_n^2, which we denote by $g_2(x)$, as follows.

$$g_2(x) = \sum_{i=0}^{\infty} F_i^2 x^i, \quad g_2(x) - x = g_2(x) - F_0^2 x^0 - F_1^2 x^1.$$

Using (11a) we have (since $F_0 = 0$)

$$\sum_{i=2}^{\infty} F_i^2 x^i = \sum_{i=2}^{\infty} (3F_{i-1}^2 - F_{i-2}^2 + 2(-1)^{i+1})x^i$$

$$= x\sum_{i=1}^{\infty} 3F_i^2 x^i - x^2 \sum_{i=0}^{\infty} F_i^2 x^i + 2x^2 \sum_{i=0}^{\infty} (-1)^i x^2$$

$$= 3xg_2(x) - x^2 g_2(x) + \frac{2x^2}{1+x}.$$

Thus

$$g_2(x)(1 - 3x + x^2) = x - \frac{2x^2}{1+x} = \frac{x - x^2}{1+x}.$$

It follows that

$$g_2(x) = \frac{x - x^2}{1 - 2x - 2x^2 + x^3}.$$

Riordan (1962) has developed generating functions for higher powers of Fibonacci numbers. For the sake of (some sort of) completeness it might be mentioned that Horadam (1965) has presented generating functions for powers of generalized Fibonacci sequences.

4. Formulae (58) and (59) enable us to derive further relationships. Those derived in Chapter III by induction could, of course, also be derived using (55), some of them even more easily. We show this on two examples, formula (13) and formula (17c).

$$L_nF_n = (\tau^n+\sigma^n)(\tau^n-\sigma^n)/\sqrt{5} = (\tau^{2n}-\sigma^{2n})/\sqrt{5} = F_{2n}. \qquad (13)$$

$$L_n^2 = (\tau^n+\sigma^n)^2 = \tau^{2n}+\sigma^{2n}+2(-1)^n = L_{2n}+2(-1)^n. \qquad (17c)$$

In many cases it is a matter of taste to choose the method by which a relationship is to be proved.

5. We show that for $n \geqslant 1$,

(a) $F_n \leqslant \tau^{n-1}$

and

(b) $F_n \geqslant \tau^{n-2}$.

Proofs.
(a)

$$F_1 = 1 = \tau^0 \quad \text{and} \quad F_2 = 1 < \tau^1.$$

Thus (a) holds for $n=1$ and for $n=2$, and hence, by induction, for all integers n.
(b) This can be proved in precisely the same manner, starting with

$$F_1 > \tau^{-1} \quad \text{and} \quad F_2 = \tau^0.$$

Next, consider

$$|F_n - \tau^n/\sqrt{5}| = |\sigma^n/\sqrt{5}| < \tfrac{1}{2} \quad \text{for } n = 0,1,2,\ldots$$

which states that F_n is the integer nearest to $\tau^n/\sqrt{5}$.

This can also be expressed by saying that F_n is the largest integer not exceeding $\tau^n/\sqrt{5}+\frac{1}{2}$. We write this

$$F_n = [\tau^n/\sqrt{5}+\tfrac{1}{2}] \quad (n = 0,1,2,\ldots) \qquad (62)$$

Similarly

$$|L_n - \tau^n| = |\sigma^n| < \tfrac{1}{2} \quad \text{for } n = 2,3,\ldots,$$

hence

$$L_n = [\tau^n+\tfrac{1}{2}] \quad (n \geqslant 2,3,\ldots) \qquad (63)$$

We have also

$$\begin{aligned}|F_{n+1}-\tau F_n| &= |\tau^{n+1}/\sqrt{5}-\sigma^{n+1}/\sqrt{5}-\tau^{n+1}/\sqrt{5}+\tau\sigma^n/\sqrt{5}| \\ &= |\sigma^n(\tau-\sigma)/\sqrt{5}| = |\sigma^n| < \tfrac{1}{2} \quad \text{for } n>1,\end{aligned}$$

which is equivalent to

$$F_{n+1} = [\tau F_n+\tfrac{1}{2}] \quad (n>1). \qquad (64)$$

Similarly,

$$|L_{n+1} - \tau L_n| = |\tau^{n+1} + \sigma^{n+1} - \tau^{n+1} - \sigma^n \tau|$$
$$= |\sigma^n(\sigma - \tau)| = |\sqrt{5}\tau^n| < \tfrac{1}{2} \quad (n > 3)$$

which is equivalent to

$$L_{n+1} = [\tau L_n + \tfrac{1}{2}] \quad (n > 3). \tag{65}$$

6. G_{m+tp}, for which we have found the expression (48) in Chapter III, was expanded into a different form by Carlitz and Ferns (1970). Starting from

$$G_{m+tp} = \alpha\tau^m\tau^{tp} + \beta\sigma^m\sigma^{tp},$$

substitute

$$\tau^t = \tau F_t + F_{t-1} \quad \text{and} \quad \sigma^t = \sigma F_t + F_{t-1}$$

from (8); this is legitimate, because τ^i and σ^i are (generalized) Fibonacci sequences.
Then

$$G_{m+tp} = \alpha\tau^m(\tau F_t + F_{t-1})^p + \beta\sigma^m(\sigma F_t + F_{t-1})^p.$$

Using the binomial theorem, we obtain

$$G_{m+tp} = \sum_{i=0}^{p} \binom{p}{i} (\alpha\tau^{i+m} + \beta\sigma^{i+m}) F_t^i F_{t-1}^{p-i},$$

that is

$$G_{m+tp} = \sum_{i=0}^{p} \binom{p}{i} F_t^i F_{t-1}^{p-i} G_{m+i}. \tag{66}$$

For $t = 2$, we find again (49).

7. We have

$$\tau^3 = 2\tau + 1, \quad \text{and} \quad \sigma^3 = 2\sigma + 1,$$

hence

$$\tau^{3n} = (2\tau)^n + \binom{n}{1}(2\tau)^{n-1} + \ldots + \binom{n}{n-1} 2\tau + 1$$

$$\sigma^{3n} = (2\sigma)^n + \binom{n}{1}(2\sigma)^{n-1} + \ldots + \binom{n}{n-1} 2\sigma + 1.$$

Addition produces

$$L_{3n} = 2^n L_n + \binom{n}{1} 2^{n-1} L_{n-1} + \ldots + \binom{n}{n-1} 2L_1 + L_0 \tag{67}$$

while subtraction, and dividing both sides by $\sqrt{5}$, produces

$$F_{3n} = 2^n F_n + \binom{n}{1} 2^{n-1} F_{n-1} + \ldots + \binom{n}{n-1} 2F_1. \tag{68}$$

8. We list a further group of formulae with binomial coefficients.

$$\sum_{i=0}^{2n} \binom{2n}{i} F_{2i} = \sum_{i=0}^{2n} \binom{2n}{i} \frac{\tau^{2i} - \sigma^{2i}}{\sqrt{5}} = \frac{1}{\sqrt{5}}[(1+\tau^2)^{2n} - (1+\sigma^2)^{2n}].$$

Using $1+\tau^2 = \tau\sqrt{5}$, $1+\sigma^2 = -\sigma\sqrt{5}$, we obtain

$$\frac{1}{\sqrt{5}}[(\tau\sqrt{5})^{2n} - (\sigma\sqrt{5})^{2n}] = 5^n F_{2n}, \quad \text{that is}$$

$$\sum_{i=0}^{2n} \binom{2n}{i} F_{2i} = 5^n F_{2n} \tag{69}$$

In precisely the same way we obtain

$$\sum_{i=0}^{2n+1} \binom{2n+1}{i} F_{2i} = 5^n L_{2n+1} \tag{70}$$

$$\sum_{i=0}^{2n} \binom{2n}{i} L_{2i} = 5^n L_{2n}, \quad \text{and} \tag{71}$$

$$\sum_{i=0}^{2n+1} \binom{2n+1}{i} L_{2i} = 5^{n+1} F_{2n+1}. \tag{72}$$

Moreover

$$\sum_{i=0}^{2n} \binom{2n}{i} F_i^2 = \frac{1}{5}\sum_{i=0}^{2n} \binom{2n}{i} (\tau^i - \sigma^i)^2$$

$$= \frac{1}{5}\left[\sum_{i=0}^{2n} \binom{2n}{i} \tau^{2i} - \sum_{i=0}^{2n} \binom{2n}{i} (-1)^i + \sum_{i=0}^{2n} \binom{2n}{i} \sigma^{2i}\right]$$

$$= \frac{1}{5}[(1+\tau^2)^{2n} - 0 + (1+\sigma^2)^{2n}] = \frac{1}{5}[(\tau\sqrt{5})^{2n} + (\sigma\sqrt{5})^{2n}]$$

that is

$$\sum_{i=0}^{2n} \binom{2n}{i} F_i^2 = 5^{n-1} L_{2n}. \tag{73}$$

Similarly

$$\sum_{i=0}^{2n+1} \binom{2n+1}{i} F_i^2 = 5^n F_{2n+1} \tag{74}$$

$$\sum_{i=0}^{2n} \binom{2n}{i} L_i^2 = 5^n L_{2n} \tag{75}$$

$$\sum_{i=0}^{2n+1} \binom{2n+1}{i} L_i^2 = 5^{n+1} F_{2n+1}. \tag{76}$$

(Some of these are mentioned in Hoggatt (1969).)

9. We are now going to derive a sum involving reciprocals of Fibonacci numbers.

$$\sum_{i=1}^{\infty} 1/F_i = 3+\sigma = 4-\tau. \tag{77}$$

This formula has been proved in a number of ways. We quote the proof in Good (1974).

Consider

$$\sum_{j=0}^{n} \frac{1}{F_{2^j}} = 3 - \frac{F_{2^n-1}}{F_{2^n}}. \tag{*}$$

This holds for $n=1$. As a consequence of (13), we can write

$$\sum_{j=0}^{n} \frac{1}{F_{2^j}} = 3 - \frac{L_{2^n} F_{2^n-1}}{L_{2^n} F_{2^n}} = 3 - \frac{L_{2^n} F_{2^n-1}}{F_{2^{n+1}}}.$$

Assume that (*) holds for all positive integers up to n. Then

$$\sum_{j=0}^{n+1} \frac{1}{F_{2^j}} = 3 - \left(\frac{L_{2^n} F_{2^n-1}}{F_{2^{n+1}}} - \frac{1}{F_{2^{n+1}}}\right).$$

To find another expression for the difference in the round brackets, we turn to (15a)

$$F_{k+m} + F_{k-m} = L_m F_k \quad \text{for even } m.$$

Set $m = 2^n$ and $k = 2^n - 1$, then

$$F_{2^{n+1}-1} + 1 = L_{2^n} . F_{2^n-1} \quad (\text{because} F_{-1} = 1).$$

It follows that

$$\sum_{j=0}^{n+1} \frac{1}{F_{2^j}} = 3 - \frac{F_{2^{n+1}-1}}{F_{2^{n+1}}}.$$

This is again (*), but n replaced by $n+1$. Thus we have proved (*) by induction.
We are, in fact, interested in

$$\lim_{n=\infty} 3 - \frac{F_{2^{n+1}-1}}{F_{2^{n+1}}}$$

which equals $3+\sigma$. (See the remark concerning Fig. III in Chapter III, according to which $F_m/F_{m-1} \rightarrow \tau$, hence $F_{m-1}/F_m \rightarrow 1/\tau = -\sigma$.) We have proved (77).

An editorial note in *The Fibonacci Quarterly*, **14** (1976), p. 187 mentions that K. Mahler has reported that

$$\sum_{i=0}^{\infty} 1/L_i \quad \text{is transcendental.}$$

V

Fibonacci series

In this chapter we present series of Fibonacci numbers which will also point to some divisibility properties, though this subject will be dealt with in greater detail in Chapter VI.

1. Let k be an odd integer. Then

$$(m+n)^k = \sum_{i=0}^{k} \binom{k}{i} m^{k-1} n^i =$$

$$= \sum_{i=0}^{(k-1)/2} \binom{k}{i} m^{k-i} n^i + \sum_{i=(k+1)/2}^{k} \binom{k}{i} m^{k-i} n^i =$$

$$= \sum_{i=0}^{(k-1)/2} \binom{k}{i} m^{k-i} n^i + \sum_{i=0}^{(k-1)/2} \binom{k}{k-i} m^i n^{k-i} =$$

$$\sum_{i=0}^{(k-1)/2} \binom{k}{i} (mn)^i (m^{k-2i} + n^{k-2i}).$$

Set $m = \tau^t$, $n = \sigma^t$, then

$$L_t^k = \sum_{i=0}^{(k-1)/2} \binom{k}{i} (-1)^{it} L_{(k-2i)t},\ k \text{ an odd integer.} \tag{78}$$

Similarly, when k is an even integer,

$$(m+n)^k = \sum_{i=0}^{(k/2)-1} \binom{k}{i} (mn)^i (m^{k-2i} + n^{k-2i}) + \binom{k}{\frac{1}{2}k} (mn)^{k/2}.$$

Set $m = \tau^t$, $n = \sigma^t$, then

$$L_t^k = \sum_{i=0}^{(k/2)-1} \binom{k}{i} (-1)^{it} L_{(k-2i)t} + \binom{k}{\frac{1}{2}k} (-1)^{tk/2}, \ k \text{ an even integer.} \quad (79)$$

Now consider, for an odd integer k

$$(m-n)^k = \sum_{i=0}^{(k-1)/2} \binom{k}{i} (mn)^i (-1)^i (m^{k-2i} - n^{k-2i}).$$

Set $m = \tau^t$, $n = \sigma^t$, then

$$(\surd 5)^k F_t^k = \sum_{i=0}^{(k-1)/2} \binom{k}{i} (-1)^{i(t+1)} \surd 5 F_{(k-2i)t}, \ k \text{ an odd integer.} \quad (80)$$

Similarly, when k is an even integer,

$$(m-n)^k = \sum_{i=0}^{(k/2)-1} \binom{k}{i} (mn)^i (-1)^i (m^{k-2i} + n^{k-2i}) + \binom{k}{\frac{1}{2}k} (mn)^{k/2} (-1)^{k/2}.$$

Set $m = \tau^t$, $n = \sigma^t$, then

$$(\surd 5)^k F_t^k = \sum_{i=0}^{(k/2)-1} \binom{k}{i} (-1)^{i(t+1)} L_{(k-2i)t} + \binom{k}{\frac{1}{2}k} (-1)^{(t+1)k/2}. \quad (81)$$

2. Formulae (78)–(81) express powers of Fibonacci numbers in terms of Fibonacci numbers. The next group of formulae expresses Fibonacci numbers in terms of powers of such numbers. They depend on the following lemma:

$$(m+n)^k = m^k + n^k - \sum_{i=1}^{[k/2]} (-1)^i (k/i) (mn)^i (m+n)^{k-2i} \binom{k-i-1}{i-1}.$$

Note that the terms on the right-hand side are integers, since for $i \leqslant k/2$

$$(k/i) \binom{k-i-1}{i-1} = \binom{k-i}{i} + \binom{k-i-1}{i-1}.$$

This lemma is mentioned in Lucas (1878), p. 209, with the remark that 'it can be proved a posteriori'. We give here such a proof.

The lemma is easily seen to hold for $k=1$, $k=2$, $k=3$. Assume that it holds for all positive integers not exceeding k. We show that then it holds for $k+1$ as well.

$$(m+n)^{k+1}=(m+n)^k(m+n)(m^k+n^k)=$$
$$=(m+n)-\sum_{i=1}^{[k/2]}(-1)^i\frac{k}{i}(mn)^i(m+n)^{k-2i+1}\binom{k-i-1}{i-1}. \qquad (*)$$

Because the lemma holds for $k-1$, we have

$$(m+n)^{k-1}=$$
$$=m^{k-1}+n^{k-1}-\sum_{i=1}^{[(k-1)/2]}(-1)^i\frac{k-1}{i}(mn)^i(m+n)^{k-2i-1}\binom{k-i-2}{i-1}$$

so that

$$mn(m^{k-1}+n^{k-1})=$$
$$=(m+n)^{k-1}mn+\sum_{i=1}^{[(k-1)/2]}(-1)^i\frac{k-1}{i}mn^{i+1}(m+n)^{k-2i-1}\binom{k-i-2}{i-1}.$$

Substituting into (*) we obtain

$$(m+n)^{k+1}=m^{k+1}+n^{k+1}+(m+n)^{k-1}mn+$$
$$+\sum_{i=1}^{[(k-1)/2]}(-1)^i\frac{k-1}{i}(mn)^{i+1}(m+n)^{k-2i-1}\binom{k-i-2}{i-1}-$$
$$-\sum_{i=1}^{[k/2]}(-1)^i\frac{k}{i}(mn)^i(m+n)^{k-2i+1}\binom{k-i-1}{i-1}.$$

The last sum (with the negative sign) $-\sum_{i=1}^{[k/2]}\ldots$ can be written

$$kmn(m+n)^{k-1} - \sum_{i=2}^{[k/2]} (-1)^i \frac{k}{i} (mn)^i (m+n)^{k-2i+1} \binom{k-i-1}{i-1} =$$

$$= kmn(m+n)^{k-1} + \sum_{i=1}^{[(k-1)/2]} (-1)^i \frac{k}{i+1} (mn)^{i+1} (m+n)^{k-2i-1} \binom{k-i-2}{i}.$$

Observing that $mn(m+n)^{k-1} + kmn(m+n)^{k-1} = (k+1),mn(m+n)^{k-1}$,

we have obtained

$$(m+n)^{k+1} = m^{k+1} + n^{k+1} + (k+1)mn(m+n)^{k-1} +$$
$$+ \sum_{i=1}^{[(k-1)/2]} (-1)^i (mn)^{i+1} (m+n)^{k-2i-1} \left[\frac{k-1}{i} \binom{k-i-2}{i-1} + \right.$$
$$\left. + \frac{k}{i+1} \binom{k-i-2}{i} \right].$$

The expression in square brackets equals

$$\frac{k+1}{i+1} \binom{k-i-1}{i},$$

therefore

$$(m+n)^{k+1} = m^{k+1} + n^{k+1} + (k+1)mn(m+n)^{k-1} +$$
$$+ \sum_{i=1}^{[(k-1)/2]} (-1)^i (mn)^{i+1} (m+n)^{k-2i-1} \frac{k+1}{i+1} \binom{k-i-1}{i}.$$

This is the same as

$$(m+n)^{k+1} = m^{k+1} + n^{k+1} - \sum_{i=1}^{(k+1)/2} (-1)^i (mn)^i (m+n)^{k-2i+1} \frac{k+1}{i} \binom{k-i}{i-1};$$

we have proved the lemma by induction.

3. Set $m=\tau^t$, $n=\sigma^t$ then

$$L_{kt}=L_t^k+\sum_{i=1}^{[k/2]}\frac{k}{i}(-1)^{i(t+1)}L_t^{k-2i}\binom{k-i-1}{i-1}. \tag{82}$$

The last term on the right-hand side will be

$2(-1)^{k(t+1)/2}$when k is even, and

when k is odd, all terms on the right-hand side are divisible by L_t. In other words:

$L_{kt}$is divisible by L_t if and only if k is odd.

Alternatively, set in the lemma $m=\tau^t$, $n=-\sigma^t$. This produces

$$F_{kt}=(\sqrt{5})^{k-1}F_t^k+\sum_{i=1}^{[(k-1)/2]}\frac{k}{i}(-1)^{it}(\sqrt{5})^{k-2i-1}\binom{k-i-1}{i-1}F_t^{k-2i} \tag{83}$$

when k is odd, and

$$L_{kt}=(\sqrt{5})^kF_t^k+\sum_{i=1}^{k/2}\frac{k}{i}(-1)^{it}(\sqrt{5})^{k-2i}\binom{k-i-1}{i-1}F_t^{k-2i} \tag{84}$$

when k is even.

4. The next group of formulae depends on the expansion of

$$\frac{m^k\pm n^k}{m\pm n}.$$

$$\frac{m^k-n^k}{m-n}=\sum_{i=0}^{(k-3)/2}(mn)^i(m^{k-2i-1}+n^{k-2i-1})+(mn)^{(k-1)/2}\text{ for odd }k\geqslant 3$$

and

$$\frac{m^k-n^k}{m-n}=\sum_{i=0}^{k/2-1}(mn)^i(m^{k-2i-1}+n^{k-2i-1})\text{ for even }k\geqslant 2.$$

Set $m = \tau^t$, $n = \sigma^t$, then

$$\frac{F_{kt}}{F_t} = \sum_{i=0}^{(k-3)/2} (-1)^{it} L_{(k-2i-1)t} + (-1)^{(k-1)t/2} \text{ for odd } k \geqslant 3 \tag{85}$$

and

$$\frac{F_{kt}}{F_t} = \sum_{i=0}^{(k/2)-1} (-1)^{it} L_{(k-2i-1)t} \text{ for even } k \geqslant 2. \tag{86}$$

(N.B. For $k = 2$ (86) reduces to (13).)

Observe that (85) and (86) show that F_{kt} is divisible by F_t, when k is even, and also when k is odd.

Consider, also

$$\frac{m^k + n^k}{m+n} = \sum_{i=0}^{(k-3)/2} (-1)^i (mn)^i (m^{k-2i-1} + n^{k-2i-1}) + (-1)^{(k-1)/2} (mn)^{(k-1)/2}$$

when k is odd, and

$$\frac{m^k - n^k}{m+n} = \sum_{i=0}^{(k/2)-1} (-1)^i (mn)^i (m^{k-2i-1} - n^{k-2i-1})$$

when k is even.

Setting $m = \tau^t$ and $n = \sigma^t$ produces, respectively

$$\frac{L_{kt}}{L_t} = \sum_{i=0}^{(k-3)/2} (-1)^{i(t+1)} L_{t(k-2i-1)} + (-1)^{(k-1)(t+1)/2} \text{ for odd } k \geqslant 3 \tag{87}$$

(we know already that L_{kt} is divisible by L_t for odd k) and

$$\frac{F_{kt}}{L_t} = \sum_{i=0}^{(k/2)-1} (-1)^{i(t+1)} F_{(k-2i-1)t} \text{ for even } k \geqslant 2. \tag{88}$$

(N.B. For $k = 2$, (88) reduces to (13), as (86) does.)

Lucas (1878) lists a number of more complex relationships, which we shall not quote here.

5. It can easily be proved by induction that

$$\frac{x}{1-x^2}+\frac{x^2}{1-x^4}+\ldots+\frac{x^{2^{n-1}}}{1-x^{2^n}}=\frac{x}{1-x}\cdot\frac{1-x^{2^n-1}}{1-x^{2^n}}.$$

Set $x=(\sigma/\tau)^r$, then

$$\frac{x}{1-x^2}=\frac{(-1)^r}{\tau^{2r}-\sigma^{2r}}=\frac{(-1)^r\sqrt{5}}{F_{2r}},$$

$$\frac{x^k}{1-x^{2k}}=\frac{(-1)^{rk}}{\tau^{2rk}-\sigma^{2rk}}=\frac{(-1)^{rk}\sqrt{5}}{F_{2rk}}, \text{ and}$$

$$\frac{1-x^{2^n-1}}{1-x^{2^n}}=\frac{F_{(2^n-1)r}}{F_{2^nr}}.$$

It follows that

$$\sum_{i=1}^{n}\frac{(-1)^{2^{i-1}r}}{F_{2ir}}=\frac{(-1)^r}{F_r}\frac{F_{(2^n-1)r}}{F_{2^nr}}. \tag{89}$$

For $r=1$, we find

$$-1+1/F_4+1/F_8+\ldots+1/F_{2^n}=-F_{2^n-1}/F_{2^n}. \tag{90}$$

6. In (58) and in (59) expand the nth powers of τ and of σ by the binomial theorem, thus:

$$F_n=\frac{1}{2^n\sqrt{5}}\left[1+n\sqrt{5}+\binom{n}{2}5+\binom{n}{3}5\sqrt{5}+\ldots\right]$$

$$-\frac{1}{2^n\sqrt{5}}\left[1-n\sqrt{5}+\binom{n}{2}5-\binom{n}{3}5\sqrt{5}+\ldots\right]$$

$$=\frac{1}{2^{n-1}}\left[\binom{n}{1}+5\binom{n}{3}+5^2\binom{n}{5}+\ldots\right] \tag{91}$$

(a formula which appears in Catalan, 1857) and

$$L_n = \frac{1}{2^n}\left[1 + n\sqrt{5} + \binom{n}{2}5 + \binom{n}{3}5\sqrt{5} + \ldots\right]$$

$$+ \frac{1}{2^n}\left[1 - n\sqrt{5} + \binom{n}{2}5 - \binom{n}{3}5\sqrt{5} + \ldots\right]$$

$$= \frac{1}{2^{n-1}}\left[1 + 5\binom{n}{2} + 5^2\binom{n}{4} + \ldots\right]. \qquad (92)$$

7. Consider $5\sum_{k=0}^{n} F^2_{r(1+k)}(-1)^{r(1+k)}$ which, by (17b), equals

$$\sum_{k=0}^{n} L_{2r(1+k)}\,(-1)^{r(1+k)} - 2\sum_{k=0}^{n}(-1)^{2r(1+k)} =$$
$$= \sum_{i=0}^{n} L_{2r(1+n-i)}(-1)^{r(1+n-i)} - 2(n+1)$$

(after changing the summation parameter from k to $n-i$).

Comparing this with (85), we find that the last expression equals

$$(-1)^{r(n+1)}F_{(2n+3)r}/F_r - 1 - 2(n+1),$$

so that

$$5\sum_{k=0}^{n} F^2_{r(1+k)}(-1)^{r(1+k)} = (-1)^{r(n+1)}F_{(2n+3)r}/F_r - 2n - 3,$$

that is

$$5\left[F_r^2(-1)^r + F_{2r}^2 + F_{3r}^2(-1)^{3r} + \ldots + F^2_{(n+1)r}(-1)^{(n+1)r} =\right.$$
$$= (-1)^{(n+1)r}F_{(2n+3)r}/F_r - 2n - 3. \qquad (93)$$

Similarly, we obtain from

$$\sum_{k=0}^{n} L_{r(1+k)}^2(-1)^{r(1+k)}$$

the series

$$L_r^2(-1)^r + L_{2r}^2(-1)^{2r} + \ldots + L_{(n+1)r}^2(-1)^{(n+1)r} =$$
$$= (-1)^{(n+1)r} F_{(2n+3)r}/F_r + 2n + 1. \tag{94}$$

More formulae of this type can be derived by exchanging k for $2k$, $3k$, . . .

8. To obtain an expression for

$$F_1^2 + F_3^2 + \ldots + F_{2n-1}^2$$

write (93) for $2n$ instead of $n+1$, and $r=1$

$$5[-F_1^2 + F_2^2 \pm \ldots + F_{2n}^2] = F_{4n+1} - 4n - 1,$$

and also for $n-1$ instead of n and $r=2$

$$5[F_2^2 + F_4^2 + \ldots + F_{2n}^2] = F_{4n+2} - 2n - 1$$

so that, by subtraction

$$5[F_1^2 + F_3^2 + \ldots + F_{2n-1}^2] = F_{4n} + 2n. \tag{95}$$

To obtain an expression for

$$L_1^2 + L_3^2 + \ldots + L_{2n-1}^2$$

write (94) for $2n$ instead of $n+1$, and $r=1$.

$$-L_1^2 + L_2^2 \pm \ldots + L_{2n}^2 = F_{4n+1} + 4n - 1$$

and also for $n-1$ instead of n, and $r=2$

$$L_2^2 + L_4^2 + \ldots + L_{2n}^2 = F_{4n+2} + 2n - 1$$

so that, again by subtraction

$$L_1^2 + L_3^2 + \ldots + L_{2n-1}^2 = F_{4n} - 2n\ . \tag{96}$$

9. To obtain sums of such expressions as (17a) and (17b), that is 'convolutions' of Fibonacci and of Lucas numbers, we prove first the formula

$$\sum_{i=0}^{n}(-1)^{i}L_{n-2i}=2F_{n+1} \tag{97}$$

This is obviously true when $n=0$ ($L_0=2$), and when $n=1$ ($L_1-L_{-1}=2$); let it be true for $n=1,2,\ldots,k$, then

$$2F_{k+1}=2F_k+2F_{k-1}=\sum_{i=0}^{k-1}(-1)^{i}L_{k-2i-1}+\sum_{i=0}^{k}(-1)^{i}L_{k-2i}=$$

$$=(-1)^{i}L_{k-2i+1}+(-1)^{k}L_{-k}=\sum_{i=0}^{k-1}(-1)^{i}L_{k-2i+1}+L_k.$$

We must show that this equals $\sum_{i=0}^{k+1}(-1)^{i}L_{k-2i+1}$; namely that

$$(-1)^{k}L_{1-k}+(-1)^{k+1}L_{-1-k}=L_k$$

or, equivalently, that $-L_{k-1}+L_{k+1}=L_k$, which we know to be true. Thus (97) is proved by induction, for all n.

Now we have (see (17b))

$$5F_iF_{n-i}=L_n-(-1)^{i}L_{n-2i}, \text{ hence}$$

$$5\sum_{i=0}^{n}F_iF_{n-i}=(n+1)L_n-2F_{n+1}=nL_n-F_n; \tag{98}$$

we also have, from (17a),

$$L_rL_{n-r}=L_n+(-1)^{r}L_{n-2r},$$

hence

$$\sum_{i=0}^{n}L_iL_{n-i}=(n+1)L_n+2F_{n+1}=(n+2)L_n+F_n\,. \tag{99}$$

If in (15b) we set $m=r$ and write $n-r$ for n, then

$$F_r L_{n-r} = F_n + (-1)^r F_{2r-n} \ .$$

$\sum_{r=0}^{n} (-1)^r F_{2r-n}$ is easily seen to be zero, therefore

$$\sum_{r=0}^{n} F_r L_{n-r} = (n+1) F_n \ . \tag{100}$$

VI

Divisibility properties

Divisibility features of Fibonacci numbers have been thoroughly studied, and we exhibit here some results.

1. If G_1 and G_2 are relatively prime, then so are any two consecutive G_i and G_{i+1}. If this were not so, then because of $G_{i-1} = G_{i+1} - G_i$, G_{i-1} would also contain any common factor of G_i and G_{i+1}, and so back to G_2 and G_1. This applies to F_i and F_{i+1}.

We see from (8), applied to $G \equiv F$, that if $F_m \equiv 0 (\text{mod } t)$ and $F_n \equiv 0 (\text{mod } t)$, then $F_{m+n} \equiv 0 (\text{mod } t)$ as well.

Also, if $F_{m+n} \equiv 0 (\text{mod } t)$ and $F_m \equiv 0 (\text{mod } t)$, then $F_{m-1}F_n \equiv 0 (\text{mod } t)$, and since F_m and F_{m-1} are relatively prime, $F_n \equiv 0 (\text{mod } t)$.

It follows, that if

$$F_i \equiv 0 (\text{mod } t) \text{ and } F_j \equiv 0 (\text{mod } t), \text{ then also}$$
$$F_{i+j} \equiv 0 (\text{mod } t) \text{ and } F_{i-j} \equiv 0 (\text{mod } t) \text{ (for } i > j).$$

In other words, the subscripts for which Fibonacci numbers are divisible by t form the positive elements of a 'module' M, say.

All such subscripts must therefore be multiples of some positive number, d say. We see this as follows:

Let n be a number in M, and let d be the smallest positive number in M. Then with an integer c, $n - cd$ must also be in M. If r is the remainder on dividing n by d, then $0 \leq r < d$, and by definition of d, we have $r = 0$. Hence n is a multiple of d. It follows that the sequence of subscripts of those Fibonacci numbers which are multiples of some number m form an arithmetic progression.

For instance, for $m = 2$, $F_3 = 2$, and F_i will be even if and only if i is a multiple of 3. Of any three consecutive Fibonacci numbers precisely one will be even.

When $m = 3$, $F_4 = 3$ and F_i will contain 3 as a factor if and only if i is a multiple of 4.

$F_6 = 8$ is the smallest Fibonacci number which contains 4, and F_i will be a multiple of 4 if and only if i is a multiple of 6. But then F_i contains 8 as well, and it follows that 4 cannot be the highest power of 2 which is a factor of any Fibonacci number.

We can also derive similar facts about Lucas numbers.

We have (6) $L_n = F_{n-1} + F_{n+1}$, and since F_{n-1} and F_{n+1} cannot both be even, L_n can only be even if both these Fibonacci numbers are odd, so that neither $n-1$ nor $n+1$ is divisible by 3. Hence n must be divisible by 3 for L_n to be even (as is true of Fibonacci numbers, see above). (We had already concluded from (24) that F_n and L_n are either both even, or both are odd.)

We show now that the product of any n consecutive terms of the Fibonacci sequence is divisible by the product of the first n terms.

This was proved by Lucas (1878) and by Carmichael (1913/14), in different ways, and the following proof is again different, though similar to that of Lucas.

Apply formula (8) to $G \equiv F$, thus:

$$F_{m+n} = F_{m-1}F_n + F_mF_{n+1}$$

and write this

$$\frac{F_{m+n}}{F_n} = F_{m-1} + \frac{F_mF_{n+1}}{F_n}.$$

It follows that

$$\frac{F_{m+n}F_{m+n-1}\ldots F_{m+1}}{F_nF_{n-1}\ldots F_1} = \frac{F_{m+n-1}\ldots F_{m+1}}{F_{n-1}\ldots F_1}\,F_{m-1} + \frac{F_{m+n-1}\ldots F_{m+1}F_m}{F_n\ldots F_1}\,F_{n+1}. \qquad (*)$$

Let us denote

$$\frac{F_tF_{t+1}\ldots F_{t+k-1}}{F_1\ldots F_k}$$

by $J(t,k)$, then (*) reads

$$J(m+1,n) = J(m+1,n-1)\ F_{m-1} + J(m,n)\ F_{n+1}$$

Clearly, $J(1,t) = 1$, and $J(t,1) = F_t$.

Now consider the triangle (similar to the Pascal triangle of binomial coefficients)

$$\begin{array}{ccccccc}
 & & & J(1,1) & & & \\
 & & J(1,2) & & J(2,1) & & \\
 & J(1,3) & & J(2,2) & & J(3,1) & \\
J(1,4) & & J(2,3) & & J(3,2) & & J(4,1)
\end{array}$$

$$- - - - - - - - - - - - - - - -$$

$$\begin{array}{ccccccc}
J(1,m+n-1) & \ldots & J(m,n) & & J(m+1,n-1) & \ldots & J(m+n-1,1) \\
J(1,m+n) & & \ldots & J(m+1,n) & \ldots & & J(m+n,1)
\end{array}$$

and so on.

All values on the left-hand edge of each row equal 1, and all those on the right-hand edge of the kth row equal F_k. These are integers.

Assume, then, that all values in the first $m+n-1$ rows are integers and consider the values in the row $m+n$. The first value is 1, the last is F_{m+n}. Any other between these two, such as $J(m+1,n)$, is obtained from the two values just above it by multiplying these, respectively, by the integers F_{m+1} and F_{n-1}, and adding. This produces again an integer, and we have proved our statement by induction. (It will be noticed that if we replace the sequence of Fibonacci numbers by the sequence of natural numbers, then we obtain the known fact that the binomial coefficients are integers.)

An analogous statement can be made about Lucas numbers, as follows.

Consider the sequence of odd-numbered Lucas numbers L_1, L_3, L_5,... . The product of any $2n-1$ consecutive terms of this sequence is divisible by the product of the first n consecutive terms.

Proof. Let

$$P_{m,n} = L_{2m-1}L_{2m+1}\ldots L_{2m-1+2(2n-2)},$$

and

$$Q_n = L_1L_3\ldots L_{2n-1}.$$

Observe that the theorem is true for $m=1$ and all n, when

$$P_{m,n}/Q_n = (L_1\ldots L_{4n-3})/(L_1\ldots L_{2n-1})$$

and also for $n=1$ and all m, when $P_{mn}/Q_n = L_{2m-1}/L_1$.

Assume that we know that it is true for $n=N-1$ and all m, and also for $n=N$ and $m=M$. Then we show that it is true for $M+1$ and N.

We have

$$P_{M+1,N} = P_{M,N} + P_{M+1,N-1}\ L_{2M+4N-5}\ (L_{2M+4N-3} - L_{2M-1}).$$

In (17a) let $n=2M+2N-2$, and $m=2N-1$ (an odd number). Then

$$L_{2M+4N-3} - L_{2M-1} = L_{2M+2N-2}\ L_{2N-1}.$$

We have now obtained

$$P_{M+1,N} = P_{M,N} + P_{M+1,N-1}L_{2N-1}L_{2M+4N-5}L_{2M+2N-2}.$$

By assumption, $P_{M+1,N-1}$ is divisible by Q_{N-1}, hence $P_{M+1,N-1}\ L_{2N-1}$ is divisible by Q_N. Thus both summands on the right-hand side are divisible by Q_N, and so is $P_{M+1,N}$. QED.

For this proof, see Conolly (1988). The proof by Carmichael (1913/14, p. 41) is different.

The following divisibility properties are proved by Cavachi (1980).

(a) $F_{kn-1} - F_{n-1}^k$ is divisible by F_n^2 $(n = 1,2,\ldots;\ k = 1,2,\ldots)$.
(b) $F_{kn-2} - (-1)^{k+1}F_{n-2}^k$ is divisible by F_n^2 $(n = 1,2,\ldots;\ k = 1,2,\ldots)$.

Proof of (a). For $k = 1$, (a) is an identity. For $k = 2$ using (11), we have

$$F_{2n-1} = F_n^2 + F_{n-1}^2,$$

so that (a) holds also for $k = 2$. Now if (a) holds for some k, then it holds also for $k + 1$, which we can see as follows:

$$F_{(k+1)n-1} = F_{kn}F_n + F_{kn-1}F_{n-1} \qquad \text{by (8).}$$

The first term on the right-hand side is divisible by F_n^2, because F_{kn} (see (85) and (86)).

The second term is congruent, by assumption, modulo F_n^2, to $F_{n-1}^k F_{n-1} = F_{n-1}^{k+1}$, so that

$$F_{(k+1)n-1} - F_{n-1}^{k+1} \text{ is divisible by } F_n^2, \text{ QED.}$$

Proof of (b). For $k = 1$, (b) is an identity. If (b) holds for some k, then it holds also for $k + 1$; we see this as follows.

Assume that (b) holds for k; that is that F_{kn-2} is congruent to $(-1)^{k+1}F_{n-2}^k$ modulo F_n^2.

Now by (8)

$$F_{(k+1)n-2} = F_{kn-1}F_n + F_{kn-2}F_{n-1}.$$

This is, by (a), which we have just proved, and by assumption, congruent, modulo F_n^2, to

$$\begin{aligned} &F_{n-1}^k F_n + (-1)^{k+1}F_{n-2}^k(F_n - F_{n-2}) \\ &= \left[F_{n-1}^k + (-1)^{k+1}F_{n-2}^k \right] F_n + (-1)^{k+2}F_{n-2}^{k+1}. \end{aligned} \qquad (*)$$

At this point we introduce a lemma:
We have

$$\frac{m^k+(-1)^{k+1}n^k}{m+n}=\sum_{i=0}^{k} m^{k-i}n^{i-1}(-1)^{i-1}.$$

If we now set $m=F_{n-1}$ and $n=F_{n-2}$, then we see that

$$F_{n-1}+(-1)^{k+1}F_{n-2}$$

is divisible by

$$F_{n-1}+F_{n-2}=F_n.$$

This lemma shows that (*) is, modulo F_n^2, congruent to $(-1)^{k+2}F_{n-2}^{k+1}$, which proves (b).

It follows from (a) and (b) that F_{nF_n} is divisible by F_n^2.

Proof. We write k for F_n.

$$F_{nk}=F_{nk-1}+F_{nk-2}\equiv\left[F_{n-1}^k+(-1)^{k+1}F_{n-2}^k\right] \pmod{k^2}.$$

Also,

$$F_{n-1}^k=(F_n-F_{n-2})^k=\sum_{i=0}^{k}\binom{k}{i}(-1)^iF_n^{k-i}F_{n-2}^i.$$

Now remember that $k=F_n$.

In the last sum the first $k-1$ terms contain, respectively, $F_n^k, F_n^{k-1},\ldots,F_n^2$, all of them divisible by F_n^2. The kth term $\binom{k}{k-1}F_n$ is also divisible by F_n^2. The last term is $(-1)^kF_{n-2}^k$.

It follows that F_{n-1}^k is congruent to $(-1)^kF_{n-2}^k \pmod{F_n^2}$ and hence F_{nk}, that is F_{nF_n}, is congruent, modulo F_n^2, to

$$(-1)^kF_{n-2}^k+(-1)^{k+1}F_{n-2}^k=0.$$

In other words, F_{nF_n} is divisible by F_n^2, QED.

Cavachi has also shown by a similar argument that

$$F_{nk},\ k=n^m$$

is divisible by F_n^{m+1} for $n=1,2,\ldots;\ m=1,2,\ldots$

2. We must deal with the question of whether any given number is a factor of some Fibonacci number. We shall prove that this is the case for ordinary Fibonacci numbers, but not for Lucas numbers.

Consider, to begin with, prime numbers. We know that there are Fibonacci numbers which contain 2 (for instance F_3), and also Fibonacci numbers which contain 5 (for instance F_5).

We consider now prime numbers different from 2 and from 5. For this purpose, we quote three facts, proved in the Appendix.

(i) $\binom{p}{n} \equiv 0 \pmod{p}$ for $1 \leqslant n \leqslant p-1$.

(ii) $\binom{p-1}{n} \equiv (-1)^n \pmod{p}$ for $1 \leqslant n \leqslant p-1$.

(iii) $\binom{p+1}{n} \equiv 0 \pmod{p}$ for $2 \leqslant n \leqslant p-1$.

From (91) we have

$$2^{p-1}F_p = \binom{p}{1} + 5\binom{p}{3} + \ldots + 5^{(p-1)/2}\binom{p}{p},$$

hence by (i), and Fermat's Theorem

$$F_p \equiv 5^{(p-1)/2} \pmod{p}.$$

It is shown in the theory of quadratic residues (see Appendix) that

$5^{(p-1)/2} \equiv 1 \pmod{p}$ if and only if p is of the form $5t \pm 1$

and

$5^{(p-1)/2} \equiv -1 \pmod{p}$ if and only if p is of the form $5t \pm 2$ where t is an integer, $(0,1,2,\ldots)$.

We have thus proved

Theorem A. If a prime p is of the form $5t \pm 1$, then $F_p \equiv 1 \pmod{p}$.
If a prime p is of the form $5t \pm 2$, then $F_p \equiv -1 \pmod{p}$.

It would be incorrect to state that, conversely, if $F_m \equiv 1 \pmod{m}$, then m must be a prime. For instance,

$$F_{22} = 17711 = 85 \times 22 + 1 \equiv 1 \pmod{22},$$

but 22 is not a prime.

Apply (91) to $n = p - 1$, then

$$2^{p-2}F_{p-1} = \binom{p-1}{1} + 5\binom{p-1}{3} + \ldots + 5^{(p-3)/2}\binom{p-1}{p-2}.$$

Hence by (ii)

$$2^{p-2}F_{p-1} \equiv -(1 + 5 + 5^2 + \ldots + 5^{(p-3)/2}) \equiv -\tfrac{1}{4}(5^{(p-1)/2} - 1) \pmod{p}.$$

Therefore if $5^{(p-1)/2} \equiv 1 \pmod{p}$, then $F_{p-1} \equiv 0 \pmod{p}$ and we have

Theorem B. If the prime p is of the form $5t \pm 1$ $(t = 0,1,\ldots)$, then

$$F_{p-1} \equiv 0 \pmod{p}.$$

On the other hand, apply (91) to $n = p + 1$. Then

$$2^pF_{p+1} = \binom{p+1}{1} + 5\binom{p+1}{3} + \ldots + 5^{(p-1)/2}\binom{p+1}{p+1},$$

hence by (iii)

$$2^pF_{p+1} \equiv 1 + 5^{(p-1)/2} \pmod{p}$$

which means that if $5^{(p-1)/2} \equiv -1 \pmod{p}$, then $F_{p+1} \equiv 0 (\mathrm{mod}\ p)$. Therefore we have

Theorem C. If the prime p is of the form $5t \pm 2$ $(t = 0,1,\ldots)$, then

$$F_{p+1} \equiv (\mathrm{mod}\ p).$$

Observe that these three theorems are in accord with formula (29). Indeed, if p is an odd prime, then either F_{p-1} or F_{p+1} is congruent to 0, modulo p, and $F_p^2 \equiv 1 (\mathrm{mod}\ p)$.

Clearly, though, $p + 1$ or $p - 1$ are not necessarily the smallest subscripts of a Fibonacci number divisible by p. For instance, $F_{18} = 2584 = 17 \times 152$ is divisible by 17, but so is the smaller number $F_9 = 17 \times 2$.

We have come to the conclusion that every prime number is the factor of some Fibonacci number. The question whether every other number is the factor of some

Fibonacci number will be answered, positively, in the next chapter.

3. We shall now derive statements analogous to Theorems A, B and C referring to Lucas numbers.

From (92) we have

$$2^{p-1}L_p = 1 + 5\binom{p}{2} + 5^2\binom{p}{4} + \ldots + 5^{(p-1)/2}\binom{p}{p-1}.$$

Hence using (i), we see

Theorem A′. $L_p \equiv 1 (\mathrm{mod}\, p)$.

$L_m \equiv 1(\mathrm{mod}\ m)$ is also possible if m is not a prime. For instance, $L_{705} \equiv 1(\mathrm{mod}\ 705)$, though 705 is not a prime number. (Hoggatt and Bicknell, 1974.)

From (92) we have also

$$2^{p-2}L_{p-1} = 1 + 5 + 5^2 + \ldots + 5^{(p-1)/2} \equiv \tfrac{1}{4}(5^{(p+1)/2} - 1)$$
$(\mathrm{mod}\ p)$.

If $5^{(p-1)/2} \equiv 1\ (\mathrm{mod}\ p)$, then $5^{(p+1)/2} \equiv 5\ (\mathrm{mod}\ p)$, hence

$$2^{p-2}L_{p-1} \equiv (5-1)/4 \equiv 1\ (\mathrm{mod}\ p) \text{ and } 2^{p-1}L_{p-1} \equiv 2 (\mathrm{mod}\ p).$$

We apply now Fermat's Theorem (see Appendix): if 2 and p are relatively prime, then $2^{p-1} \equiv 1(\mathrm{mod}\ p)$. There follows

Theorem B′. If the prime p is of the form $5t \pm 1$ (t integer), then $L_{p-1} \equiv 2\ (\mathrm{mod}\ p)$.

Once more from (92) we have

$$2^p L_{p+1} = 1 + 5\binom{p+1}{2} + 5^2\binom{p+1}{4} + \ \ldots\ + 5^{(p+1)/2}\binom{p+1}{p+1}.$$

Applying (iii), this means

$$2^p L_{p+1} \equiv (1 + 5^{(p+1)/2})\ (\mathrm{mod}\ p).$$

We use again Fermat's Theorem to obtain

$$2^p L_{p+1} \equiv (1 + 5 \times 5^{(p-1)/2})\ (\mathrm{mod}\ p)$$

which means, since $5^{(p-1)/2} \equiv -1 (\mathrm{mod}\ p)$,

Theorem C′. if the prime p is of the form $5t \pm 2$ (t integer), then $L_{p+1} \equiv -2 \pmod p$.

4. We derive yet another congruence from (A′) and (82).
When t is an odd prime, then (82) can be written

$$L_{kt} = L_t^k + \sum_{i=1}^{[k/2]} \frac{k}{i} \binom{k-i-1}{i-1} L_t^{t-2i}$$

and for $t=1$

$$L_k = L_1^k + \sum_{i=1}^{[k/2]} \frac{k}{i} \binom{k-i-1}{i-1} L_1^{k-2i}$$

Because $L_1 = 1$, and $L_p \equiv 1 \pmod p$, it follows that $L_{kt} \equiv L_k \pmod t$ when t is an odd prime.

As a matter of fact $L_{2k} \equiv L_k \pmod 2$ is also true. We know that L_k is even if k is divisible by 3. Then $2k$ is also divisible by 3, and L_{2k} is even. On the other hand, if L_k is odd, then either $k \equiv 1 \pmod 3$ or $k \equiv 2 \pmod 3$, hence either $2k \equiv 2 \pmod 3$ or $2k \equiv 1 \pmod 3$ and L_{2k} is odd.

5. If follows from (83) that if an odd prime is contained in F_t precisely v times, then F_{pt} contains p precisely $(v+1)$ times, and $F_{p^m t}$ contains p precisely $(v+m)$ times. For $p=5$ this means that the highest power of 5 contained in F_r is the same as the highest power of 5 contained in the subscript r. It also follows that no F_i can be a power of 5, except $F_5 = 5$, because $F_d = 5^a$ $(a > 1)$ would mean that d contains 5^a, but d cannot be larger than F_d itself when $d > 5$.

Now consider $p=2$. We have seen, as a consequence of (24), that F_n and L_n cannot have any common divisor larger than 2. Let F_n be divisible by 2. Then as we have seen, L_n is also divisible by 2. If, then, 2^a $(a > 1)$ is the highest power of 2 contained in F_n, then L_n will contain just 2, and no higher power of 2. On the other hand, if the highest power contained in F_n is 2, then L_n will contain 4, but no higher power of 2. We can see this as follows:

We have seen that 8 is not a factor of any Lucas number, hence L_n cannot be divisible by 2^n $(n > 3)$ either. But L_n cannot contain just 2 (and no higher power), because if it did, then we would have

$$L_n = 2a \text{ with } (2;a) = 1 \text{ and } F_n = 2b \text{ with } (2;b) = 1$$

where $(m;n)$ is our notation for the largest common divisor of m and n. Hence, by (24),

$$4(5b^2 - a^2) = \pm 4 \text{ that is } 5b^2 - a^2 = \pm 1$$

which is impossible, since a as well as b is odd.

Formula (24) also provides information about prime factors of Fibonacci and of Lucas numbers.

When n is odd, then (24) can be written

$$4 \equiv -L_n^2 \pmod{5F_n^2}$$

and this means that -1 is a quadratic residue modulo those primes which divide F_n.

Now (-1) is a quadratic residue modulo 2, and modulo those primes which have the form $4t+1$. Therefore odd primes which divide F_n with odd n are of the form $4t+1$.

Again, when n is odd, then (24) can be written

$$4 \equiv 5F_n \pmod{L_n^2}$$

which says that 5 is a quadratic residue modulo those primes which divide L_n^2.

Now 5 is a quadratic residue modulo those primes which are of the form $10t \pm 1$, and hence these are the primes which divide L_n for odd n.

When n is even, then

$$4 \equiv -5F_n^2 \pmod{L_n^2}.$$

Now $\left(\frac{5}{p}\right) = 1$ when $p = 10k \pm 1$, and $\left(\frac{-1}{p}\right) = 1$ when $p = 4k+1$, therefore $\left(\frac{-5}{p}\right) = 1$ if p is simultaneously of these two forms, that is p has the form

$$20t+1, \text{ or } 20t+9.$$

Also, $\left(\frac{-5}{p}\right) = -1$ when $p = 10n \pm 3$, and $\left(\frac{-1}{p}\right) = -1$ when $p = 4k+3$ therefore $\left(\frac{-5}{p}\right) = 1$ if p is of the form $20t+3$, or $20t+7$.

Consequently, the odd primes which divide L_n when n is even are of the form $20t + 1,3,7$, or 9.

6. We turn to another topic: divisibility of a Fibonacci number by another Fibonacci number. This will be prsented in three theorems.

Theorem I. If s is divisible by t, then F_s is divisible by F_t.

We have already seen that this is true, from formulae (85) and (86). Here we give a proof on more classical lines.

It follows from (8), when $G \equiv F, m = t, n = rt$, that if F_{rt} is divisible by F_t, then so is

$F_{(r+1)t}$.

Now F_{2t} is divisible by F_t (see formula (13)) and hence, by induction, F_{kt} is divisible by F_t for any integer k.

As a matter of fact, this theorem holds for any U_n as defined in Chapter II, section 2 (see Siebeck, 1846).

Theorem II. If $(s;t) = d$, then $(F_s;F_t) = F_d$.

(a;b) denotes the largest common factor of a and b.

Let $s > t$. We find the largest common factor of s and t by the Euclidean algorithm, as follows:

$$\begin{aligned}
&s = p_0t + r_1 \ (0 < r_1 < t)\\
&t = p_1r_1 + r_2 \ (0 < r_2 < r_1)\\
&r_1 = p_2r_2 + r_3 \ (0 < r_3 < r_2)\\
&\vdots\\
&r_{z-2} = p_{z-1}r_{z-1} + r_z \ (0 < r_z < r)\\
&r_{z-1} = p_zr_z
\end{aligned}$$

and $r_z = (s;t) = d$.

Now consider

$$(F_s;F_t) = (F_{p_0t+r_1};F_t)$$

which equals, by (8)

$$(F_{r_1}F_{p_ot-1} + F_{r_i+1}F_{p_0t};F_t).$$

Because F_t divides F_{p_0t} by Theorem I, the last expression equals

$$(F_{r_1}F_{p_0t-1};F_t).$$

Because F_{p_0t-1} and F_{p_0t}, and hence also F_{p_0t-1} and F_t are relatively prime, we have obtained

$$(F_s;F_t) = (F_{r_1};F_t).$$

Continuing in parallel with the steps of the Euclidean algorithm, we reach

$$(F_{r_2};F_{r_1}) = \ldots = (F_{r_z};F_{r_{z-1}}) = F_d.$$

This completes the proof of the theorem.

Theorem III. If F_s is divisible by F_t $(t \neq 2)$, then s is divisible by t. If the assumption holds, then $(F_s;F_t) = F_t$. By Theorem II, the fact that F_t equals $F_{(s;t)}$ implies that s is divisible by t.

(We had to stipulate $t \neq 2$, because F_s is divisible by $F_2 = 1$ in any case.)

Theorem IV. If $(F_s;F_t) = F_d$, then $(s;t) = d$.

If $(F_s;F_t) = F_d$, then both F_s and F_t are divisible by F_d.

Hence, by Theorem III, both s and t are divisible by d. But d is, in fact, the largest common divisor (not just some divisor), because if the largest common divisor were rd, say, and $r > 1$, then (by Theorem II), (F_s,F_t) would be F_{rd}, and not F_d, as we have assumed.

This completes the proof of Theorem IV.

7. It follows from Theorem I, that if s is a prime, then the factors of F_s cannot be Fibonacci numbers. However, F_s may have (other) factors. For instance, 19 is a prime, and $F_{19} = 4181 = 113 \times 37$. This leads naturally to the following question: if s is not a prime, can F_s be a prime?

Let s be divisible by v, and let v be divisible by w. The F_s is divisible by F_v, and also by F_w. Therefore, if s is not a prime, then F_s can only be a prime if F_v as well as F_w equal 1, that is v as well as w can only be 1 or 2, and s can only be 1,2, or 4. Hence if s is not a prime, then F_s can only be a prime for $F_4 = 3$.

If $(s;t) = 1$, or $(s;t) = 2$, then $(F_s;F_t) = 1$, so that F_{st} is divided by F_s, by F_t and by F_sF_t as well.

As a matter of fact, F_{st} is also divided by the product $F_s.F_t$ when $(s;t) = 5$. This is a consequence of the fact that the highest power of 5 contained in F_r is also the highest power of 5 contained in r (see the remark above in section 5). Here is the proof:

Let $(s;t) = 5$, say $s = 5a$, $t = 5^k b$, with $(5;a) = (5;b) = (a;b) = 1$. Then $(F_s;F_t) = F_{(s;t)} = 5$, F_{st} contains F_s as well as F_t, and also 5^{k+1}. Hence F_{st} contains F_sF_t as well.

Conversely, if F_{st} is divided by F_sF_t, then $(s;t)$ must be 1, or 2, or 5.

Proof. (D. Jarden, 1946). let F_u $(u > 0)$ be the smallest Fibonacci number containing the prime p. The subscript u is called by Lucas the rank of apparition of p, and we know that it is a factor of, or equal to $p - 1$ or $p + 1$.

Let p^a be the highest power of p dividing F_u, and let $r = p^b k$, p and k being relatively prime. Then, as we have seen, the highest power of p dividing F_r is $p^{a+b+g(r)}$. We must introduce $g(r)$, because of the exceptional role played by the prime 2, when the power b equals 1. We define $g(r) = 1$ when $p^b = 2$, and $g(r) = 0$ otherwise. Clearly $g(rs) \leqslant g(r) + g(s)$.

Now suppose that $(s;t)$ differs from 1,2, and 5. This will lead to a contradiction.

We have seen that $F_{(s;t)}$ cannot be a power of 5 when $(s;t) \neq 5$. We have assumed that $(s;t) > 2$, hence $F_{(s;t)} > (s;t)$, so that there exists a prime $p \neq 5$, which divides $F_{(s;t)}$.

The rank of apparition of p, u say, is a divisor of $p - 1$, or of $p + 1$, and therefore relatively prime to p.

Write

$$(s;t) = uh,\ s = s'(s;t) = s'uh,\ t = t'(s;t) = t'uh,\ st = s't'u^2h^2.$$

If p^v and p^w are the highest powers of p dividing s and t respectively, then the highest power of p dividing

$$F_s \text{ is } p^{v+a+g(s)}$$
$$F_t \text{ is } p^{w+a+g(t)}$$
$$F_{st} \text{ is } p^{v+w+a+g(st)}$$
$$F_sF_t \text{ is } p^{v+w+2a+g(s)+g(t)}.$$

We are studying the case when F_{st} is divisible by F_sF_t, that is when

$$v+w+a+g(st) \geq v+w+2a+g(s)+g(t)$$

that is

$$a \leq g(st)-g(s)-g(t) \leq 0$$

which is absurd. Hence $(s;t)$ must be 1, or 2, or 5.

8. Consider the case when L_{kt}/L_t is an integer. We know already that then k must be odd. We shall now prove that if L_k is odd, then L_{kt}/L_t is also odd.

L_t, the denominator, may be (a) odd, or (b) even. We treat these two cases separately.

(a)
L_t is odd. This is so if and only if t is not a multiple of 3. Because if we assume that L_k is odd, k cannot be a multiple of 3 either, and neither can kt. L_{kt} will, as a consequence, be odd, and the ratio L_{kt}/L_t, being the ratio of two odd numbers, will also be odd.

(b)
L_t is even. The subscript t must be a multiple of 3, kt is also a multiple of 3, and L_{kt} is even, as L_t is. We must therefore prove, that for odd k, L_t and L_{kt} contain the same highest power of 2.

We know already that such a highest power cannot be different from 2 or 4. Now if the highest power in L_t is 4, then L_{kt} will also contain 4, and no Lucas number can contain a higher power of 2. Therefore on division the powers of 2 cancel out. If the highest power in L_t is 2, then by (82) this is also the highest power contained in L_{kt}. Hence the ratio L_{kt}/L_t is, in either case, odd. (For the gist of this proof, see Carmichael 1913/14, pp. 36–39.)

The converse does not hold. When L_k is even, then L_{kt}/L_t may be even, or odd. Examples:

$$L_3 = 4.$$
$$L_{15}/L_5 = 1364/11 = 124$$
$$L_{18}/L_6 = 5778/18 = 321.$$

9. Theorem II in section 6 above referred to Fibonacci numbers. We shall now prove an analogous theorem for Lucas numbers.

$$\text{If } (s;t) = d, \text{ then } (L_s;L_t) = L_d$$

provided that both s/d and t/d are odd integers.

Proof. We know that under the proviso mentioned, L_d is a common divisor of L_s and L_t. (See the remark after formula (82).) Now we have to prove that it is the *largest* common divisor.

By (13) $F_{2s} = F_sL_s$ and $F_{2t} = F_tL_t$, so that $(L_s;L_t)$ is a common divisor of F_{2d} by Theorem I in section 6), since by assumption $(2s;2t) = 2d$.

It follows that $(F_{2s};F_{2t}) = F_{2d}$, that is that F_{2d} is a multiple of $(L_s;L_t)$ and thus L_dF_d is a multiple of $(L_s;L_t)$.

If the largest factor that F_d has in common with L_s and L_t is m, say, then $(L_s;L_t) = L_dm$, and we have to show that $m = 1$.

We know from (24) that L_s and F_s have no common factor larger than 2, and this holds also for L_t and F_tF_d is a factor of F_s and of F_t, so that we can also say that

L_s and F_d (and also L_t and F_d) have no common factor larger than 2.

If in both cases the common factor were 2, then we would have $(L_s,L_t) = 2L_d$. We must therefore show that L_s/L_d and L_t/L_d cannot both be even.

Now the ratios s/d and t/d can not both be divisible by 3, since if they were, then $3d$ would be a larger common factor of s and t than d is. Therefore at least one of $L_{s/d}$ and $L_{t/d}$ must be odd (see the statement about divisibility by 3 at the beginning of this chapter) and hence (see section 8) L_s/L_d or L_t/L_d (or both) must be odd. This completes the proof.

10. On a slightly different topic, we prove the theorem: When a Fibonacci number is being divided by another Fibonacci number, and the remainder equals r, say, then either r is itself a Fibonacci number, or it is the difference between two Fibonacci numbers.

Proof. Let F_{n+m} be divided by F_n. From (10a) we have

$$F_{n+m} - L_mF_n = (-1)^{m+1}F_{n-m}.$$

If F_{n-m} is still larger than F_n, repeat the process, and divide F_{n-m} by F_n.

If, at the final stage, $(-1)^{m+1}F_{n-m}$ is negative, because m is even, while F_{n-m} is positive, then apply

$$F_{n+m} - (L_m - 1)\ F_n = (F_n - F_{n-m}).$$

However, if $(-1)^{m+1}F_{n-m}$ is negative, because m is odd, but F_{n-m} is negative, then apply

$$F_{n+m} - (L_m - 1)F_n = (F_n - F_{m-n})$$

(Observe that if F_{n-m} is negative, then F_{m-n} is also a Fibonacci number.)

The same theorem holds when we replace F by L. Again, if L_{n-m} is negative, then L_{m-n} is also a Lucas number. For other generalized Fibonacci numbers this is not true.

VII

Congruences and uniformity

1. Consider Fibonacci sequences and the smallest residues of the terms modulo some integer m. Denote the residues of F_n by $R_n(m)$.

For instance, when $m = 5$, we find

n	0	1	2	3	4	5	6	7	8	9	10	11	12	13	14	15	16	17	18	19	20	21
$R_n(5)$	0	1	1	2	3	0	3	3	1	4	0	4	4	3	2	1	2	2	4	1	0	1

Observe that this sequence repeats after 20 terms. It is easy to see that any such sequence, modulo any number, must repeat periodically. Looking at successive pairs, such as in this example 01, 11, 12, 23, 30, . . . we know that there are not more than $m^2 - 1$ of them which are distinct, ignoring the trivial pair 00. The first pair which reappears will be the one we started with, in the example 01, because if R_iR_{i+1} turns up a second time, then this happens because the pair before it was $R_{i-1}R_i$, and so on back to the start.

It must be pointed out that such an argument fails when the mechanism is $u_{n+2} = au_{n+1} + bu_n$ and b is not relatively prime to m. For instance, take $u_{n+2} = u_{n+1} + 2u_n$, modulo 4. Starting with 01, we generate

0 1 1 3 1 3 1 3 . . .

The repeated pair is 13, but we did not start with it. The reason becomes clear when we notice that the 'backward' congruence

$$u_{n+2} - u_{n+1} = 2x \pmod 4$$

has two solutions for x, viz. 3, but also 1. The pair 13 can be preceded by 3, but also by 1.

We call the string of terms between repetitions a cycle and we denote the period of a sequence of generalized Fibonacci numbers G_i modulo m by $P(m, G)$.

A cycle finishes when a pair reappears. An individual number may, of course, reappear before the end of the cycle. In particular, if $R_n(m)$ appears again after k entries, that is if $F_{n+k} \pmod m$ equals $F_n \pmod m$, then k must be a factor of the period, as we have seen in the last chapter. Hence it follows that when $R_0(m) = 0$, then

Theorem I.

$F_{P(m,F)}$ is divisible by m.

Such an argument does not apply to Lucas numbers, because L_0 does not equal 0.

We shall derive a number of results concerning the periods of smallest residues. For convenience, we exhibit the following table.

m	2	3	4	5	6	7	8	9	10	11	12	13	14	15
$P(m,F)$	3	8	6	20	24	16	12	24	60	10	24	28	48	40
$P(m,L)$	3	8	6	4	24	16	12	24	12	10	24	28	48	8

m	16	17	18	19	20	21	22	23	24	25	26	27	28	29	30
$P(m,F)$	24	36	24	18	60	16	30	48	24	100	84	72	48	14	120
$P(m,L)$	24	36	24	18	12	16	30	48	24	20	84	72	48	14	24

Observe that all periods, except those for $m = 2$, are even. We show why this must be so by assuming that a period is odd, and deducing from this assumption that m must be 2. We follow closely the proof given in Wall (1960). It applies to $P(m, F)$, for which we write more briefly k.

Suppose k to be odd, equal to $2x + 1$. Then, starting at both ends and proceeding towards the middle, we find (all congruences are modulo m)

$$-F_k \equiv 0 = F_0\ ,\quad F_{k-1} \equiv 1 = F_1\ ,\quad -F_{k-2} = F_k - F_{k-1} \equiv F_0 + F_1 = F_2$$

$$\vdots$$

$$(-1)^{t-1}F_{k-t} = (-1)^{t-1}F_{k-t+2} + (-1)^t F_{k-t+1} \equiv F_{t-2} + F_{t-1} = F_t$$

$$\vdots$$

$$(-1)^{x-2}F_{x+2} \equiv F_{x-1}\ ,\qquad (-1)^{x-1}F_{x+1} \equiv F_x\ .$$

From the last congruence it follows that

$$F_{x-1} \equiv 0 \quad \text{if } x \text{ is odd, and}$$
$$F_{x+2} \equiv 0 \quad \text{if } x \text{ is even.}$$

If x is even, then it follows from the last but one congruence again that $F_{x-1} \equiv 0$.

Now we know that the smallest subscript in the Fibonacci sequence for which the residue is 0, say the subscript d, is a factor of $x - 1$, and hence of $2x - 2$, and also that it is a factor of k, that is of $2x + 1$. Consequently d is a factor of

$$2x + 1 - 2x + 2 = 3\ .$$

It follows that, because $F_3 = 2 \equiv 0(\text{mod } m)$, m must be 2, QED.

An analogous proof applies to Lucus numbers, but not to all generalized Fibanacci numbers.

2. Let n be a factor of m. The cycle of smallest residues modulo m repeats after

$cP(n, G)$ terms, where c is an integer. Hence
Theorem II. If n divides m, then $P(n,G)$ divides $P)m,G)$.

If m is of the form

$$\prod_i p_{p_i}^{e_i}$$

where the p_i are distinct primes and the e_i are positive integers, then $P(p_i^{e_i},G)$ divides $P(m,G)$ it follows that
Theorem III. $P(m,G)$ is the least common multiple of $P(p_i^{e_i},G)$.

For instance, 10 is the least common multiple of 2 and 5. Therefore the last decimal digit of a Fibonacci number is repeated after 60 terms (a remark made by Lagrange), and that of a Lucas number after 12 terms.

We notice also in our table that if p is a prime different from 5, then $P(p, F)$ is a factor of $p^2 - 1$. This must be so, because when $p \neq 5$, then either, F_{p-1} *or* F_{p+1} is a multiple of p (Theorems B and C in the previous Chapter) so that, as a consequence of Theorem I in that Chapter, $F_{p^2-1} \equiv 0(\text{mod } p)$. No period can be longer than $p^2 - 1$, and it must therefore be a factor of, or equal to $p^2 - 1$.

Conolly (1981), pp. 217–272, contains a different proof.

It follows from (8) that if $P(m, F) = k$, then the residues of any generalized sequence G_i will also repeat after k terms, so that $P(m, G)$ divides $P(m, F)$, or is equal to it. Hence

Theorem IV. The longest possible period of any generalized Fibonacci sequence is that of the ordinary Fibonacci sequence.

If the period $P(m, F)$ is different from, that is smaller than $m^2 - 1$, then there must be some pairs (d_i, d_{i+1}) which have not appeared in the cycle starting with $(0, 1)$. We can then construct a sequence with seed (d_i, d_{i+1}) in which all pairs will be different from that with seed $(0, 1)$.

For instance, the sequence modulo 5 with which we started this Chapter does not contain $(2, 1)$, the seed of the Lucas sequence. The smallest residues modulo 5 are then

2 1 3 4 .

We have now accounted for all $5^2 - 1 = 24$ possible pairs.

However, we must not conclude that the Lucas sequence is always missing from the Fibonacci sequence of residues. When $m = 6$, then the Fibonacci sequence of period 24

0 1 1 2 3 5 2 1 3 4 1 5 0 5 5 4 3 1 4 5 3 2 5 1

contains the Lucus sequence, though not all possible pairs. The other sequences are

0 2 2 4 1 4 4 2 (period 8)

and

0 3 3 (period 3).

3. We shall now deal with theorems concerning the lengths of periods of ordinary Fibonacci sequences.

Theorem I. If the prime p is of the form $5t \pm 1$ (t an integer), then $P(p, F)$ divides $p - 1$.

Proof. We have seen in Chapter VI, section 2, that if p has the form $5t \pm 1$, then $F_p \equiv 1 (\text{mod}\, p)$ and that $F_{p-1} \equiv 0 \,(\text{mod}\, p)$. This implies that the period is $p - 1$, or a factor of $p - 1$.

The smallest prime of the form $5t \pm 1$, for which the period is actually smaller than $p - 1$ is $p = 29$, with period 14.

Theorem II. If the prime p is of the form $5t \pm 2$, then $P(p, F)$ divides $2p + 2$.

Proof. We have seen that in this case $F_{p+1} \equiv 0 \,(\text{mod}\, p)$, and that $F_p \equiv -1 \,(\text{mod}\, p)$. It follows that $F_{p+2} \equiv -1 \,(\text{mod}\, p)$, so that the period does not equal $p + 1$.

From (8) we have

$$F_{2p+3} = F_p F_{p+2} + F_{p+1} F_{p+3} \,,$$

therefore $F_{2p+3} \equiv 1$ (*mod* p). Together with $F_{2p+2} \equiv 0$ (mod p) this proves the theorem.

The smallest prime of the type we consider. for which the period does not equal $2p + 2$, but a factor of it, is $p = 47$, with period 32.

4. We know that a period modulo m can not be larger than $m^2 - 1$. This lead us to the question of whether there are, in fact, values of m such that the period *equals* $m^2 - 1$.

We shall now prove that this is only possible for $m = 2$, and for $m = 3$. We find by inspection that it is indeed the case for these two values.

Modulo 2	0 1 1	period 3
Modulo 3	0 1 1 2 0 2 2 1	period 8.

In order to prove that these are the only cases where the period modulo m equals $m^2 - 1$, we shall first consider primes, then powers of primes, and finally other composite moduli.

(a)

m is a prime number.

We have dealt with $m = 2$; $P(5, F) = 20$, which is not equal to $5^2 - 1$. Prime numbers which differ from 2 and from 5 have either the form $5t \pm 1$, or the form $5t \pm 2$.

($a1$) Let $p = 5t \pm 1$. Theorem I states that the period is at most $p - 1$, and therefore not $p^2 - 1$.

(a2) Let $p = 5t \pm 2$. By Theorem II the period is at most $2p + 2$, and this equals $p^2 - 1$ only when $p = 3$.

(b)

We turn to powers of primes.

In this case we use the proposition that if t is the largest integer with $P(p^t, F) = P(p, F)$, then $P(p^s, F) = p^{s-}\ P(p, F)$ for $s > t$.

The proof of this proposition is sketched in Wall (1960) and, differently, in Chang (1986). We refer the reader to these sources, but we prove the proposition here by elementary methods for $p = 2$, and for $p = 5$, following Kramer and Hoggatt (1973).

We have

$$P(2, F) = 3 \text{ and } P(4, F) = 6 \neq 3.$$

Also,

$$P(5, F) = 20 \text{ and } P(25, F) = 100 \neq 20.$$

Hence we have to prove that

$$P(2^n, F) = 2^{n-1}3, \qquad \text{and} \qquad P(5^n, F) = 5^{n-1}20 = 5^n 4.$$

Concerning $p = 2$, we must prove that

(i) $\qquad F_{2^{n-1}3} \equiv F_0 \pmod{2^n}$

and that

(ii) $\qquad F_{(2^{n-1}3)+1} \equiv F_1 \pmod{2^n}$.

We have $F_3 \equiv 0 \pmod 2$ and assume, to start the proof by induction, that

$$F_{2^{n-1}3} \equiv 0 \pmod{2^n}, \qquad (n = 1, 2, \ldots, m)$$

Now

$$F_{2^m 3} = F_{3(2^{m-1})} L_{3(2^{m-1})} \qquad \text{(formula 13)}$$

The second factor is divisible by 2, hence $F_{2^m 3}$ is divisible by 2^{m+1}. This proves (i). Also, from (11) we have

$$F_{(2^m 3)+1} = F^2_{(2^{m-1}3)+1} + F^2_{2^{m-1}3} \ .$$

We study the two terms on the right-hand side separately.

We have assumed that

$$F_{2^{m-1}3} \equiv 0 \pmod{2^m}$$

hence

$$F^2_{2^{m-1}3} \equiv 0 \pmod{2^m} .$$

From this, and from (20a), that is

$$F^2_{(2^{m-1}3)+1} = F_{(2^{m-1}3)+2} F_{2^{m-1}3} + (-1)^{2^{m-2}3)+2}$$

we obtain

$$F^2_{(2^{m-1}3)+1} \equiv 1 \pmod{2^m} .$$

Adding,

(ii) $$F_{(2^m3)+1} \equiv 1 \pmod{2^m}$$

which completes the proof for $p = 2$.

Concerning $p = 5$, we have to prove that

$$F_{5^n4} \equiv 0 \pmod{5^n}$$

and that

$$F_{(5^n4)+1} \equiv 1 \pmod{5^n} .$$

From (83) we know that

$$F_{5^n4} \equiv 0 \pmod{5^n} \quad \text{(Chapter VI, section 5)}$$

and since F_{5^n4} contains F_{5^n} as a factor, it follows that

$$F_{5^n4} \equiv 0 \pmod{5^n} .$$

Also

$$F_{(5^n4)+1} = F^2_{(5^n2)+1} + F^2_{(5^n2)} .$$

The second term on the right-hand side is $\equiv 0 \pmod{5^n}$, and from this fact and (20a), that is

$$F^2_{(5^n2)+1} = F_{5^n2} F_{(5^n2)+2} + (-1)^{(5^n2)+2} \equiv 1 \pmod{5^n}$$

the required result follows.

The proposition, applied to any prime p, proves that

$$P(p^s, F) \leqslant p^{s-1}(p^2 - 1)$$

which, in turn, is less than $p^{2s} - 1$ for $s >$.

(c)

We have still to deal with composite moduli, which are not prime powers.

If $m = rs$, where r and s are powers of different primes, then $P(rs, F)$ is not larger than the least common multiple of $r^2 - 1$ and $s^2 - 1$, and this is less than $r^2s^2 - 1$.

If m is a product of more than two prime powers, then an analogous argument applies.

We have thus proved that 2 and 3 are the only moduli for which $P(m,F) = m^2 - 1$.

5. The present section will deal with relationships between $P(m,F)$ and $P(m,G)$, where G_i is any generalized Fibonacci sequence. We shall quote proofs from Wall (1960).

First, we present cases where $P(m,F) = P(m,G)$ for any G.

Theorem V. If a prime p is of the form $5t \pm 2$, then

$$P(p^s,F) = P(p^s,G)$$

provided G_0, G_1 and p are mutually relatively prime.

Proof. Let $P(p,G) = h$, say. Then

$$G_h - G_0 \equiv 0 \pmod p \qquad \text{and} \qquad G_{h+1} - G_1 \equiv 0 \pmod p \ .$$

Using (8), this can be written

$$\begin{aligned} G_1F_h + G_0(F_{h-1} - 1) &\equiv 0 \pmod p \\ G_2F_h + G_1(F_{h-1} - 1) &\equiv 0 \pmod p. \end{aligned}$$

We show that the determinant of this system of two linear congruences, that is

$$D = G_1^2 - G_0G_2 = G_1^2 - G_0G_1 - G_0^2$$

is not congruent to 0 modulo p, so that the only possible solution is

$$F_h \equiv 0 \pmod p \ , \qquad F_{h-1} \equiv 1 \pmod p \ .$$

In other words, the period of F, say k, is a factor of h. But we know that in fact h is a factor of k (see Theorem IV in section 2), therefore h must equal k.

So we have to show that $D \not\equiv 0 \pmod p$.

If D were congruent to 0 modulo p, then

$$4G_0^2 + 4G_0G_1 + G_1^2 = (2G_0 + G_1)^2 \equiv 5G_1^2 \qquad \pmod p \ .$$

But by the assumption G_1^2 cannot be a multiple of p. Therefore $D \equiv 0 \pmod p$ would imply that 5 is a quadratic residue of p. However, if p is of the form $5t \pm 2$, this is not the case, and we have proved Theorem V.

Wall (1960) has also proved the following theorems:

(1) If $m = 5^t$, then $P(m,F) = P(m,G)$ when $G_1^2 - G_0G_2$ is not divisible by 5 and $P(m,F) = 5P(m,G)$ when $G_1^2 - G_0G_2$ is divisible by 5.

Observe that the second case holds when G_i is the Lucas sequence.

(2) If $m = p^t$, $p > 2$ and if $P(m,G)$ is odd, then $P(m,F)$ is odd, then $P(m,F) = 2P(m,G)$.

(3) If $m = p^t, p > 2$ and if $P(m, F) = 4t + 2$, then there exists a sequence G_i for which $P(m, G) = 2t + 1$.
(4) If $m = p^t, p > 2$ and $\neq 5$, and if $P(m, G)$ is even, then it equals $P(m, F)$.

6. Many complex problems in probability and in statistics are studied by the experimental method of simulation. An example is the Needle Problem of Buffon: imagine a needle of known length to be thrown 'at random' onto a large board with parallel lines drawn on it. The probability of the needle hitting a line can be described by a formula which contains π. The probability of the crossing can be estimated by many random drops and by counting how often a line has been crossed. From this estimate, the value of π can be estimated.

Such an experiment can be simulated without having to use needles and a board, by drawing random numbers for the distance of the centre of a needle with given length from the lines, and for its inclination.

The sequence of random numbers can be generated within a computer in various ways. Of course, truly random sequences cannot be obtained by a predictable mechanical process, but 'pseudo-random' numbers can be produced, which satisfy to some degree tests of randomness, appropriate to the problem to be solved.

It has been found that some sequences generated by Fibonacci-like mechanisms are often acceptable as being pseudo-random.

Thus Miller and Prentice (1968) have looked at the mechanism

$$u_{n+3} = u_{n+1} + u_n \; .$$

Conolly (1981) has made an extensive study of the more general mechanism

$$u_N = u_{n+1} + u_n \qquad (N \geqslant n + 2)$$

with various values of N and various moduli, using the seed $(0, \ldots, 0, 1)$, which contains $N - 1$ zeroes and one 1. He has shown that such a seed produces the longest possible period for given N and m. (For $N = n + 2$ we know this already, see Theorem IV in section 2.)

We quote here from Conolly (1981) an extract of his Table 8.2 which lists the periods for various N and m

$N-n$	m				
	2	3	4	5	10
2	3	8	6	20	60
3	7	13	14	24	168
5	21	121	42	781	16 401
10	889	1640	1778	1 953 124	1 736 327 236
12	3255	6560	6510	13 563 368	203 450 520*

* This is the case favoured by Conolly.

As an example, we choose $N - n = 5$ and $m = 4$. The sequence is

0 0 0 0 1 0 0 0 1 1 0 0 1 2 1 0 1 3 3 1 1
0 2 0 2 1 2 2 2 3 3 0 0 1 2 3 0 1 3 1 3 1

7. A useful random collection of numbers will usually be expected to contain each of 0, 1, 2, ... If this is not so, the cycle will be called defective. The smallest modulus for which the Fibonacci sequence is defective is $m = 11$. The period is 10, and the cycle

0,1,1,2,3,5,8,2,10,1

does not contain 4, 6, 7, or 9.

In the Tribonacci case

$$u_{n+3} = u_{n+2} + u_{n+1} + u_n$$

and seed (0,0,1), the smallest modulus which leads to a defective cycle is $m = 8$.

0 0 1 1 2 4 7 5 0 4 1 5 2 0 7 1

does not contain either 3 or 6.

Even when all values $0, 1, \ldots, m-1$ appear in the cycle, they may not all appear with the same frequency. However, for close similarity with truly random numbers one would wish all smallest residues to appear equally often in a cycle. If this happens, we call the cycle 'uniform'.

Let us therefore see which moduli, if any, lead to uniformity in Fibonacci sequences.

Clearly, for a cycle to be uniform, the period must be divisible by the modulus. We start our study by considering prime moduli.

(i) $p = 2$. The cycle 0 1 1 is not uniform. Anyway, $2^2 - 1$ is not divisible by 2.
(ii) Let the prime p be of the form $5t \pm 1$. Then, as we have seen, the period divides $p - 1$, and cannot be divisible by p.
(iii) Let the prime p be of the form $5t \pm 2$. We have seen that the period divides, $2p + 2$, which again is not divisible by p.
(iv) The remaining case is that of $p = 5$. Its cycle was exhibited at the beginning of this chapter. We find, by inspection, that the cycle is uniform.

Kuipers and Shiua (1972) prove (ii) and (iii) in a way different from our argument.

We turn to composite moduli.

By Theorem II above, $P(n, F)$ divides $P(m, F)$ if n divides m. It is easy to see that if the cycle modulo m is uniform, then so is the cycle modulo n. Instead of giving a formal proof, we give an example which makes the reason clear.

Modulo 12, the period equals 24.

0 1 1 2 3 5 8 1 9 10 7 5 0 5 5 10 3 1 4 5 9 2 11 1

Modulo 3, the period equals 8.

0 1 1 2 0 2 2 1 ;0 1 1 2 0 2 2 1 ;0 1 1 20 2 2 1

Modulo 4, the period equals 6.

0 1 1 2 3 1 ;0 1 1 2 3 1 ;0 1 1 2 3 1 ;0 1 1 2 3 1

The numbers in the cycle for modulus 12 are congruent modulo 3 to those corresponding in the modulo 3 cycle, and they are congruent modulo 4 to those in the modulo 4 cycle, though the first are not smallest residues modulo 3 or 4.

Of course, these cycles are not uniform. But if the top cycle were uniform, it would follow that the lower ones were also uniform.

Consequently no m which contains as a factor any prime apart from 5 can lead to a uniform distribution.

It remains for us to consider moduli which are powers of 5. We prove

Theorem VI. The least residues modulo 5^t of a Fibonacci sequence are uniformly distributed for all integer $t = 1, 2, \ldots$ (Niederreiter, 1972.).

Proof. We know that $P(5^k, F) = 4(5^k)$.

To say that the smallest residues modulo 5^k appear equally often in a cycle is equivalent to saying that each of them appears precisely four times. We prove the equivalent fact that each of them appears at most four times.

We know this to be the case for $k = 1$. Assume that for some $k \geqslant 2$, the congruence

$$F_n \equiv a \pmod{5^{k-1}}$$

has precisely four solutions n_1, n_2, n_3, n_4 for any given a, $(0 \leqslant n_i \leqslant 4\,(5^{k-1}) - 1)$.

For instance, let $k = 2$ and $a = 3$, then the following Fibonacci numbers are congruent to 3, modulo, 5:

$$F_4 = 3\ , \qquad F_6 = 8\ , \qquad F_7 = 13\ , \qquad F_{13} = 233\ ,$$

that is the n_i are 3, 6, 7 and 13.

Now let n_0 be a solution of the congruence

$$F_n \equiv a \pmod{5^k}$$

and $0 \leqslant n_0 \leqslant 4(5^k) - 1$.

Then also

$$F_{n_0} \equiv a \pmod{5^{k-1}}\ ,$$

hence n_0 is congruent, modulo 5^{k-1}, to one of the n_i $(i = 1, 2, 3, 4)$.

For the reader's convenience, we continue our example by exhibiting the cycle of period 100 of the Fibonacci sequence modulo 25.

0 1 1 2 3 5 8 13 21 9 5 14 19 8 2 10 12 22 9 6 15 21 11 7 18
0 18 18 11 4 15 19 9 3 12 15 2 17 19 11 5 16 21 12 8 20 3 23 1 24
0 24 24 23 22 20 17 12 4 16 20 11 6 17 23 15 13 3 16 19 10 4 14 18 7
0 7 7 14 21 10 6 16 22 13 10 23 8 6 14 20 9 4 13 17 5 22 2 24 1

In this list 3 appears as the smallest residue of F_4, F_{33}, F_{46} *and* F_{67}.

The subscripts 4, 33, 46 and 67 are indeed congruent, modulo 5, to the subscripts which we have quoted above, 4, 13, 6 and 7, in this order.

We must still prove that each of the four n_i leads to one single value n_0, so that those four n_i cannot lead to more than four values n_0.

Suppose, to the contrary, that

$$F_m \equiv a \pmod{5^k} \text{ and that } m \neq n \ (0 \leqslant m \leqslant 4(5^k) - 1)$$

but

$$m \equiv n \pmod{4(5^{k-1})},$$

so that $F_n \equiv F_m \pmod{5^k}$.

From (91), we have

$$\sum_{j=0} 5^j \binom{n}{2j+1} \equiv 2^{n-m} \sum_{j=0} 5^j \binom{m}{2j+1} \pmod{5^k}$$

and because

$$2^{4(5^{k-1})} \equiv 1 \pmod{5^k} \qquad \text{(the Fermat–Euler Theorem)}$$

$$\sum_{j=0} 5^j \left[\binom{n}{2j+1} - \binom{m}{2j+1}\right] \equiv 0 \pmod{5^k}. \tag{*}$$

On the left-hand side use

$$\begin{aligned}\binom{n}{2j+1} &= \sum_{i=0}^{2j+1} (n_i - m) \binom{m}{2j+1-1} \\ &= \binom{m}{2j+1} + \sum_{i=1}^{2j+1} \binom{n-m}{i} \binom{m}{2j+1-i},\end{aligned}$$

so that

$$5^j \left[\binom{n}{2j+1} - \binom{m}{2j+1}\right] = 5^j \sum_{i=1}^{2j+1} \binom{n-m}{i} \binom{m}{2j+1-i}. \tag{**}$$

It is easy to see that cancelling out 5s from $i!$ against 5^j, when $i \leqslant 2j+1$, leaves at least one power of 5 uncancelled. There is also a factor of 5^{k-1} in $n-m$, so that each term in (**) is divisible by 5^k. Therefore there remains in (*) a term with $j=0$, that is

$$n-m \equiv 0 \pmod{5^k} \ .$$

We know already that $n-m \equiv 0 \pmod{4(5^{k-1})}$, therefore

$$n-m \equiv 0 \ (\mathrm{mod}\ 4(5^k) \ , \qquad \text{or} \qquad m=n \ .$$

8. In the last section we studied uniformity in Fibonacci sequences. In the present section, we study general second order sequences.

Narkiewicz (1984) writes:

> Necessary and sufficient conditions for a linear recurrent sequence of order k to be uniformly distributed are known only in the case 2, 3 and 4.

He deals only with the case $k = 2$, and we shall follow his example. We shall quote results without the lengthy proofs, but we shall illustrate them.

Consider the mechanism

$$u_{n+2} = au_{n+1} + bu_n$$

where a, b, u_0 and u_1 are integers.

Let $f(x) = x^2 - ax - b = (x-s)(x-t)$ with discriminant a^2+4b. Then u_n is uniformly distributed modulo a prime m if and only if

(i) p does not divide $b(u_1 - su_0)$ when s and t are rational, or
(ii) p does not divide $st(u_1 - su_0)(u_1 - tu_0)$ when s and t are not rational.

Moreover, u_n is uniformly distributed modulo m if and only if it is uniformly distributed modulo all primes which divide m, provided none of the following holds:

(a) 4 divides m, and a is not congruent 2 (mod 4);
(b) 4 divides m, and b is not congruent 3 (mod 4);
(c) 9 divides m, and $a^2 + b$ is not divisible by 9.

Examples.
(1) The Fibonacci mechanism has $a = b = 1$, $a^2+4b = 5$. Hence the Fibonacci sequence is uniform modulo any power of 5, but modulo no other integer. On the other hand, in the Lucas sequence $u_0 = 2$ and $st(1-2s)(1-2t) = 5$, which is divisible by 5, hence the Lucus sequence is not uniform modulo any integer m.

(2) $a = 6$, $b = -9$. The discriminant is zero, the double root is 3, a rational integer. Any prime divides 0, so the uniformity depends on the seed (u_0, u_1). For uniformity to hold, p must not divide $-9(u_1 - 3u_0)$.

Let the seed be $(2,5)$. $\qquad -9(5-6) = 9$.

The sequence modulo 3 (which divides 0) is not uniform:

2 2 0 0 0

but the sequence modulo 5 is uniform

2 0 2 2 4 1 0 1 1 2 3 0 3 3 1 4 0 4 4 3 .

(3) $a = 4$, $b = 1$. The discriminant is 20, and the roots are the irrational numbers $2+\sqrt{5}$ and $2-\sqrt{5}$.

The primes which divide 20 are 2 and 5.

$$(2+\sqrt{5})(2-\sqrt{5})(u_1-(2+\sqrt{5})u_0)(u_1-(2-\sqrt{5})u_0) = u_0^2 + 4u_0u_1 - u_1^2 = v \text{ ,say.}$$

Again, uniformity depends on the seed.

When the seed is $(0, 1)$, then $v = -1$, not divisible by either 2 or by 5.

Modulo 2:

0 1 0 1 0 . . .

is uniform.

Modulo 5

0 1 4 2 2 0 2 3 4 4 0 4 1 3 3 0 3 2 1 1

is also uniform.

On the other hand, when the seed is $(1,2)$, hence $v = 5$, then the sequence modulo 2

1 0 1 0

is uniform, but the sequence modulo 5

1 2 4 3

is not.

For modulo 10 we find, when the seed is $(0, 1)$, that

0 1 4 7 2 5 2 3 4 9 0 9 6 3 8 5 8 7 6 1

is uniform, but when the seed is $(1,2)$, then

1 2 9 8

is not uniform.

VIII

Continued fractions and convergents

1. Consider the ratio G_{n+1}/G_n. It equals $(G_n + G_{n-1})/G_n$, and this can be written

$$1+\frac{1}{G_n/G_{n-1}}.$$

We deal with G_n/G_{n-1} as we dealt with G_{n+1}/G_n, and obtain

$$1+\cfrac{1}{1+\cfrac{1}{G_{n-1}/G_{n-2}}}$$

and, continuing in this manner, we reach eventually

$$1+\cfrac{1}{1+\cfrac{1}{1+\cfrac{1}{1+\cfrac{1}{1+\cfrac{}{\ddots 1+\cfrac{1}{G_2/G_1}}}}}}.$$

What we have done is to expand the ratio G_{n+1}/G_n into a 'continued fraction'.
When $G_i \equiv F_i$, the we obtain the continuous fractions

$$F_2/F_1 = 1, \quad F_3/F_2 = 1+\frac{1}{1}, \quad F_4/F_3 = 1+\cfrac{1}{1+\cfrac{1}{1}}$$

and so on. These are the initial portion of the expansion of F_{n+1}/F_n into a continued fraction, and are called the 'convergents' of that fraction.

In order to find the largest common factor of F_{n+1} and F_n (we know that it equals

1, because two successive Fibonacci numbers are relatively prime) the Euclidean algorithm can be described by writing

$$F_{n+1}/F_n = 1+\frac{1}{1+F_n/F_{n-1}} = 1+\frac{1}{1+\frac{F_{n-1}}{F_{n-2}}} = 1 = \cfrac{1}{1+\cfrac{1}{1+\ddots\, 1}}$$

2. The theory of continued fractions is extensive (see the Appendix) and we shall quote results which refer to those continued fractions which we have just mentioned.

As n increases without bounds, the continued fractions of the convergents tend to an infinite continued fraction

$$1+\cfrac{1}{1+\cfrac{1}{1+\cfrac{1}{1+\ddots}}}$$

$= z$, say, We have $z = 1 + 1/z$, that is $z^2 - z - 1 = 0$. Since z is positive, we find

$$z = \tfrac{1}{2}(1+\sqrt{5}) = \tau.$$

We have found

$$\lim_{n=\infty} F_{n+1}/F_n = \tau. \qquad (101)$$

This follows also from

$$\lim_{n=\infty} \frac{\tau^{n+1}-\sigma^{n+1}}{\tau^n-\sigma^n}$$

because $|\tau| > 1$ and $|\sigma| < 1$. Formula (101) is a special case of a formula in Chapter II, section 3.

It follows from (101) that

$$\lim_{n=\infty} F_n/F_{n-m} = \lim_{n=\infty} \frac{F_n}{F_{n-1}}\frac{F_{n-1}}{F_{n-2}}\cdots\frac{F_{n-m+1}}{F_{n-m}} = \tau^m. \qquad (101a)$$

Clearly, the infinite series

$$\frac{F_2}{F_1} + \left(\frac{F_3}{F_2} - \frac{F_2}{F_1}\right) + \ldots + \left(\frac{F_{n+1}}{F_n} - \frac{F_n}{F_{n-1}}\right) + \ldots$$

also converges to τ. Using (29), this can be written

$$1 + \sum_{n=2}^{\infty} \frac{(-1)^n}{F_n F_{n-1}} = \tau. \tag{102}$$

To show the speed of convergence, and also to exemplify other features which we shall mention later, we give here a list of the first convergents of the infinite continued fraction.

n	1	2	3	4	5	6	7	8	9
$\frac{F_{n+1}}{F_n}$ (to 3 decimals)	1	2	1.5	1.667	1.6	1.625	1.619	1.61	1.618

(τ equals 1.618 to 3 decimals' accuracy.)

3. The speed of convergence of a sequence can be increased by applying 'Aitken acceleration', which relies on the fact that

$$x_1^*, x_2^*, \ldots, x_n^*$$

converges more rapidly than

$$x_1, x_2, \ldots, x_n$$

when

$$x_n^* = \frac{x_{n+r}x_{n-r} - x_n^2}{x_{n+r} - 2x_n + x_{n-r}}.$$

We show how this works for the convergence of the sequence $x_n = F_{n+1}/F_n$ (Phillips, 1984).

In this case the numerator of x_n^*, N_n say, reads

$$\frac{F_{n+r+1}}{F_{n+r}}\frac{F_{n-r+1}}{F_{n-r}} - \frac{F_{n+1}^2}{F_n^2} = \frac{(F_{n+r+1}F_{n-r+1} - F_{n+1}^2)F_n^2 - (F_{n+r}F_{n-r} - F_n^2)F_{n+1}^2}{F_{n+r}F_{n-r}F_n^2}$$

$$= \frac{(-1)^{n+r}F_r^2(F_n^2 + F_{n+1}^2)}{F_{n+r}F_{n-r}F_n^2}$$

by (20a), $h = -k = 1$
that is

$$N_n = \frac{(-1)^{n+r}F_r^2 F_{2n+1}}{F_{n+r}F_{n-r}F_n^2} \quad \text{by (11).}$$

We look now at the denominator of x_n^*, at D_n, say.

We have

$$D_n = \frac{F_{n+r+1}}{F_{n+r}} - \frac{2F_{n+1}}{F_n} + \frac{F_{n-r+1}}{F_{n-r}}$$

$$= \frac{(F_{n+r+1}F_n - F_{n+1}F_{n+r})F_{n-r} - (F_{n+1}F_{n-r} - F_{n-r+1}F_n)F_{n+r}}{F_{n+r}F_{n-r}F_n}$$

$$= \frac{(-1)^{n+r}F_r(F_{n+r} + (-1)^{r-1}F_{n-r})}{F_{n+r}F_{n-r}F_n} \quad \text{by (20a).}$$

As a consequence of (10b), we can write

$$D_n = \frac{(-1)^{n+r}F_r^2(F_{n+1} + F_{n-1})}{F_{n+r}F_{n-r}F_n}.$$

We have obtained

$$x_n^* = \frac{N_n}{D_n} = \frac{(-1)^{n+r}F_r^2 F_{2n+1}}{F_n(-1)^{n+r}(F_{n+1} + F_{n-1})F_r^2)}$$

and this means, using (13)

$$x_n^* = \frac{F_{2n+1}}{F_{2n}}.$$

Indeed, F_3/F_2, F_5/F_4, F_7/F_6, ... converges more quickly to τ than F_2/F_1, F_3/F_2, F_4/F_3, F_5/F_4, ... does.

4. We shall now investigate the manner in which the convergents approach their limit.

(i) First, we show that

$$F_{n+1}/F_n < F_{n-1}/F_{n-2} \quad \text{when } n \text{ is even, and that}$$

$$F_{n+1}/F_n > F_{n-1}/F_{n-2} \quad \text{when } n \text{ is odd.} \tag{103a}$$

These inequalities are special consequences of (20a). Let $h = 1$, and $k = -2$, then

$$F_{n+1}F_{n-2} - F_nF_{n-1} = (-1)^n F_1 F_{-2} = (-1)^{n+1}.$$

Division by F_nF_{n-2}, which is positive, establishes the required result.

(ii) Next, we show that

$$\tau - \frac{F_{n+1}}{F_n} \quad \text{and} \quad \tau - \frac{F_n}{F_{n-1}}$$

have different signs. From (29)

$$(-1)^n = F_{n+1}F_{n-1} - F_n^2$$
$$= (F_{n-1} + F_n\tau)(F_{n+1} - F_n\tau) \quad (\text{because } \tau^2 = 1 + \tau)$$

so that

$$F_{n+1} - F_n\tau = \frac{(-1)^n}{F_{n-1} + F_n\tau}. \tag{103b}$$

Since $F_{n-1} + F_n\tau > 0$, it follows that F_{n+1}/F_n is smaller than τ when n is odd, and larger than τ when n is even.

(iii) Also

$$\left|\tau - \frac{F_{n+1}}{F_n}\right| < \left|\tau - \frac{F_n}{F_{n-1}}\right|. \tag{103c}$$

This follows at once from (103b), because the denominator on the right-hand side increases with n.

(iv) If in (103b) we divide both sides by F_n and take absolute values, then we have

$$\left|\tau - \frac{F_{n+1}}{F_n}\right| = \left|\frac{1}{F_n^2(\tau + F_{n-1}/F_n)}\right| < \frac{1}{F_n^2}. \tag{104}$$

We shall now prove a refinement of this inequality, namely the theorem: of any two consecutive convergents, at least one satisfies

$$\left|\tau - \frac{F_{n+1}}{F_n}\right| < \frac{1}{\sqrt{5}F_n^2}.$$

From (104) we see that the theorem will be proved when we have shown that for any n at least one of the values

$$\tau + \frac{F_n}{F_{n+1}} \quad \text{and} \quad \tau + \frac{F_{n-1}}{F_n} \quad \text{exceeds } \sqrt{5}.$$

Now $\sqrt{5} = \tau + 1/\tau$, so that it is sufficient to show that at least one of the two ratios F_{n-1}/F_n and F_n/F_{n+1} exceeds $1/\tau$. But we know already that F_n/F_{n-1} and F_{n+1}/F_n are alternately larger and smaller than τ. It follows that F_{n-1}/F_n and F_n/F_{n+1} are alternately smaller and larger than $1/\tau$, which proves the theorem.

N.B. In the general case when other values than τ are expanded into a continued fraction, the theorem does not hold generally. It remains valid, if we replace 'of any two' by the phrase 'of any three' (see Perron 1929, pp. 48ff).

It would be incorrect to replace $\sqrt{5}$ by $c > \sqrt{5}$, because for large enough n it would not be true that $\tau + F_n/F_{n+1} > c$. Indeed, let $c = \sqrt{5} + a$, $a > 0$. For sufficiently large n, $F_n/F_{n+1} < 1/\tau + a$, and then

$$\tau + F_n/F_{n+1} < \tau + \frac{1}{\tau} + a = \sqrt{5} + a = c.$$

5. We consider briefly the ratios of generalized Fibonacci numbers. G_{n+1}/G_n converges also to τ with increasing n, because

$$\frac{G_{n+1}}{G_n} - \frac{F_{n+1}}{F_n} = \frac{G_{n+1}F_n - G_nF_{n+1}}{G_nF_n}$$

$$= (-1)^{n+1}G_0/G_nF_n$$

by (21), and this ratio converges, with increasing n, to 0.

6. In (102), we have expressed τ as an infinite series. We shall express it now as an infinite product. We have

$$\lim_{n=\infty} \frac{F_2}{F_1} \times \frac{F_3/F_2}{F_2/F_1} \times \ldots\ldots \times \frac{F_{n+1}/F_n}{F_n/F_{n-1}} = \lim_{n=\infty} \frac{F_{n+1}}{F_n} = \tau.$$

But

$$\frac{F_{i+2}/F_{i+1}}{F_{i+1}/F_i} = \frac{F_{i+2}F_i}{F_{i+1}^2} = \frac{F_{i+1}^2 + (-1)^{i+1}}{F_{i+1}^2}$$

by (29), so that

$$\tau = \prod_{i=1}^{\infty} \left(1 + \frac{(-1)^{i+1}}{F_{i+1}^2}\right). \tag{105}$$

7. We derive a further relationship between Fibonacci numbers and Lucas numbers, in the form of a continued fraction.

In formula (15a) let $n = tm$, then

$$\frac{F_{(t+1)m}}{F_{tm}} = L_m - \frac{(-1)^m}{F_{tm}/F_{(t-1)m}}.$$

Developing further, we obtain

$$\frac{F_{(t+1)m}}{F_{tm}} = L_m - \cfrac{(-1)^m}{L_m - \cfrac{(-1)^m}{F_{(t-1)m}/F_{(t-2)m}}}$$

and so on, until L_m has appeared t times. Then we have

$$\frac{F_{(t+1)m}}{F_{tm}} = L_m - \cfrac{(-1)^m}{L_m - \cfrac{(-1)^m}{L_m - \ddots \cfrac{(-1)^m}{L_m}}}. \tag{106}$$

With increasing t, this expression tends to τ^m.

Examples: $m = 3$ ($L_3 = 4$)

$$t = 3\ldots 144/34 = 4+\cfrac{1}{4+\cfrac{1}{4}};$$

$$t = 4\ldots 610/144 = 4+\cfrac{1}{4+\cfrac{1}{4+\cfrac{1}{4}}}$$

IX

Fibonacci representation

1.

Theorem I. Any positive integer N can be expressed as a sum of distinct Fibonacci numbers:

$$N = F_{k_1} + F_{k_2} + \ldots + F_{k_r}$$

such that

(i) $\quad k_{i+1} \leqslant k_i - 2 \qquad (i = 1,2,\ldots,r-1)$

and

(ii) $\quad k_r \geqslant 2.$

This is called Zeckendorf's Theorem.

When N is itself a Fibonacci number (for instance, $N = 1$, or 2, or 3) then the theorem is trivial. We see by inspection that it is also true for $N = 4 = F_4 + F_2$. Assume it to be true for all integers not exceeding F_n, and let $F_{n+1} > N > F_n$.

Now $N = F_n + (N - F_n)$; $N - F_n < F_n$, so that, by our assumption, $N - F_n$ can be expressed in such a way that (i) and (ii) hold.

We call a representation of N as a sum of Fibonacci numbers for which (i) and (ii) hold 'canonical'. About such a representation we prove

Theorem II. The canonical representation is unique. (Lekkerkerker, 1952.)

First, we show that in the canonical representation of N, when $F_{n+1} > N > F_n$, the number F_n must appear. If it did not, then no sum of smaller Fibonacci numbers, obeying (i) and (ii), could add up to N. If n is even, then this follows from $n = 2k$, say,

$$F_{2k-1} + F_{2k-3} + \ldots + F_3 = F_{2k} - F_1 = F_n - 1, \quad \text{(see 34))}$$

and if n is odd, say $2k+1$, then it follows from

$$F_{2k}+F_{2k-2}+\ldots+F_2=F_{2k+1}-F_1=F_n-1 \quad \text{(see (35))}.$$

Again, in the representation of $N-F_n$ the largest Fibonacci number not exceeding $N-F_n$ must appear, and we see that it cannot be F_{n-1}. Thus Theorem II is proved by induction.

In the same way it can be proved that in Theorems I and II the reference to Fibonacci numbers can be replaced by a reference to Lucas numbers, provided (ii) is replaced by

(ii′) if $k_0=1$, then $k_1=k_2=0$.

Indeed, $k_0=k_1=1$ is unnecessary, because, $L_0+L_1=L_2$, and $k_0=k_2=1$ is unnecessary, because $L_0+L_2=L_1+L_3$. (Carlitz *et al.*, 1972b).

Of course, non-canonical representations by sums of Fibonacci numbers are also possible. For instance,

$$25=F_8+F_4+F_2 \quad \text{(canonical)}$$

but also

$$25=F_7+F_6+F_4+F_2\ .$$

However, provided that any Fibonacci number is to be used not more than once, we have

Theorem III. If and only if N has the form F_n-1, then the canonical representation is the only possible one.

If $N=F_n-1$, then F_{n-1} must be in the representation, because for $n\geqslant 3$

$$\sum_{i=2}^{n-2} F_i=F_n-2 \quad \text{(see (33))}$$

and would therefore not be large enough to make up F_n-1.

It remains then to make up $(F_n-1)-F_{n-1}=F_{n-2}-1$, and by the previous argument we must have F_{n-3} in the representation. So we carry on and find

$$F_n-1=F_{n-1}+F_{n-3}+\ldots+F_r$$

where r is either 2 or 3.

Conversely, if we know that only one representation is possible, then N must have the form given. This will now be proved.

The smallest Fibonacci number appearing in the representation must be F_2 or F_3. (We ignore the possibility of using F_1, since it can always be replaced by F_2.) If it were F_r and $r>3$, then another valid representation could be constructed, replacing F_r by $F_{r-1}+F_{r-2}$. Moreover, the difference between the subscripts can not be larger than 2, because if we had, for instance, F_t+F_{t-s} $(s>2)$, then again F_t could be replaced by $F_{t-1}+F_{t-2}$. Therefore, for the representation to be unique, we must have either

$$F_{2r}+F_{2r-2}+\ldots+F_2 \text{ which, by (35), equals } F_{2r+1}-1$$

or

$$F_{2r+1}+F_{2r-1}+\ldots+F_3 \text{ which, by (34), equals } F_{2r+2}-1 \ .$$

Carlitz (1968) has computed the number of possible representations, canonical or non-canonical, for other cases as well.

2. We prove now two properties of the canonical representation.

Let $F_{n+1}>N>F_n$, and

$$N=F_{k_1}+\ldots+F_{k_r} \ .$$

Property 1. For any two numbers in the sum, say F_{k_i} and F_{k_j}, with $F_{k_i}>F_{k_j}$, we have $F_{k_i}>2F_{k_j}$.

Proof. The difference between k_i and k_j is at least 2, and

$$F_i=F_{i-1}+F_{i-2}>2F_{i-2} \ .$$

Property 2. This relates to F_{k_r} and to $D=F_{n+1}-N$. We prove that $F_{k_r}<2D$.

There may be summands apart from F_{k_1} and F_{k_r}. They would be, respectively, $\leqslant F_{k_1-2}$, $\leqslant F_{k_1-4}$, $\ldots$ $\leqslant F_{k_1-2j}$, because the gap between subscripts of two summands is at least 2. Now

$$\begin{aligned} D &\geqslant F_{k_1+1}-F_{k_1}-F_{k_1-2}-\ldots-F_{k_1-2j}-F_{k_r} \\ &=F_{k_1-1}-F_{k_1-2}-\ldots-F_{k_1-2j}-F_{k_r} \\ &=F_{k_1-3}-\ldots-F_{k_1-2j}-F_{k_r} \\ &\ \vdots \\ &=F_{k_1-2j-1}-F_{k_r}\geqslant F_{k_1-2j-1}-F_{k_1-2j-2}=F_{k_1-2j-3} \ . \end{aligned}$$

Moreover,

$$2F_{k_1-2j-3} > F_{k_1-2j-2} \geqslant F_{k_r} \ ,$$

so that

$$2D > F_{k_r} \quad \text{QED}$$

Properties 1 and 2 will be useful in the analysis of Fibonacci Nim, a game which we describe in the next chapter.

3. Let a representation of $N-1$ be

$$N-1 = F_{k_1} + \ldots + F_{k_r} \quad (k_r \geqslant 2), \text{ that is}$$
$$N = F_{k_1} + \ldots + F_{k_r} + F_1 \ .$$

Define

$$A(N-1) = F_{k_1+1} + \ldots + F_{k_r+1}$$

and

$$B(N-1) = F_{k_1+2} + \ldots + F_{k_r+2} + F_1 \ .$$

Then

$$N + A(N-1) = B(N-1) \ .$$

Define also

(i) $\qquad a(N) = A(N-1) + 1 = F_{k_1+1} + \ldots + F_{k_r+1} + F_2 \ ,$

(ii) $\qquad b(N) = B(N-1) + 1 = F_{k_1+2} + \ldots + F_{k_r+2} + F_3.$

A change from $b(N)$ to $a(N)$ will be called a 'shift'. A shift of a canonical, or a non-canonical representation of a number produces again the same, shifted, number, in canonical or non-canonical form. To prove this, we have to show that if

$$F_{k_1} + \ldots + F_{k_r} = F_{j_1} + \ldots + F_{j_s} = M, \text{ say with}$$
$$k_1 > k_2 > \ \ldots > k_r \geqslant 2 \quad \text{and} \quad j_1 > j_2 > \ldots > j_s \geqslant 2 \ ,$$

then

$$F_{k_1-1} + \ldots + F_{k_r-1} = F_{j_1-1} + \ldots + F_{j_s-1} \ .$$

Clearly, this is (trivially) true for $M=1$. Let it be true for all integers not exceeding $M-1$.

Now either $k_1=j_1$, or $k_1 \neq j_1$, say $k_1 > j_1$.

If $k_1=j_1$, then $F_{k_2}+\ldots+F_{k_r}=F_{j_2}+\ldots+F_{j_s}<M$, and the theorem is proved by induction.

If $k_1>j_1$, then we must have $j_1=k_1-1$, because if we had $F_{j_1}=F_{k_1-2}$ (or less), then even adding all Fibonacci numbers less than F_{j_1} would produce only (or less than)

$$F_{k_1-2}+\ldots+F_2=F_{k_1}-2<M, \text{ by (33).}$$

Let then $j_1=k_1-1$.

Now consider k_2. We have $k_2=k_1-1$, or $k_2=k_1-2$, or $k_2<k_1-2$.

If k_2 were equal to k_1-1, which is equal to j_1, then the theorem follows as in the previous case for $k_1=j_1$.

If $j_1=k_1-1$, and $k_2=k_1-2$, then

$$M=F_{k_1}+F_{k_2}+\ldots+F_{k_r}=F_{k_1-1}+F_{j_2}+\ldots+F_{j_s} \quad ,$$

that is

$$F_{k_1}-F_{k_1-1}+F_{k_2}+\ldots+F_{k_r}=F_{j_2}+\ldots+F_{j_s} \quad ,$$

or

$$2F_{k_2}+F_{k_3}+\ldots+F_{k_r}=F_{j_2}+\ldots+F_{j_s} \qquad (*)$$

where

$$j_2<j_1=k_1-1, \text{ hence } j_2 \leqslant k_1-2=k_2 \quad .$$

However, $j_2<k_2$ is impossible, because it would mean that (by (33))

$$F_{j_2}+F_{j_3}+\ldots+F_{j_s} \leqslant F_2+F_3+\ldots+F_{k_2-1}<F_{k_2+1}<2F_{k_2}$$

which contradicts (*). Hence $j_2=k_2$, and (*) becomes

$$F_{k_2}+F_{k_3}+\ldots+F_{k_r}=F_{j_3}+\ldots+F_{j_s}<M$$

and, by assumption

$$F_{k_2-1}+\ldots+F_{k_r-1}=F_{j_1-1}+\ldots+F_{j_s-1}\ . \qquad (**)$$

But

$$F_{k_1-1}=F_{j_1}=F_{j_1-1}+F_{j_1-2}=F_{j_1-1}+F_{j_2-1}$$

so that, adding F_{k_1-1} on the left-hand side and $F_{j_1-1}+F_{j_2-1}$ on the right-hand side of (**), the theorem is again proved.

We must still deal with the case $F_{k_2}<F_{k_1-2}$. In this case

$$\begin{aligned} F_{j_2}+F_{j_3}+\ldots+F_{j_s} &= F_{k_1}+\ldots+F_{k_r}-F_{k_1-1} \\ &= F_{k_1-2}+F_{k_2}+\ldots+F_{k_r} \\ &= M', \text{ say, where } M'<M. \end{aligned}$$

Therefore, by assumption,

$$F_{j_2-1}+F_{j_3-1}+\ldots+F_{j_s-1}=F_{k_1-3}+F_{k_2-1}+\ldots+F_{k_r-1}$$

and the proof is complete.

Thus $a(N)$ is a shift of $b(N)$, and N that of $a(N)$. (The proof is taken from Carlitz, 1968).

4. In the next chapter we shall make use of the following three characteristics of N, $a(N)$, and $b(N)$.

(C_1) $a(N)$ as well as $b(N)$ form monotone increasing sequences.
(C_2) $b(N)-a(N)=N$.
(C_3) Each positive integer is either an $a(N)$, or a $b(N)$, but never both at the same time.

(C_1 and C_2) follow at once from the definitions.

In the definition (i) of $a(N)$ the last term is F_2, and if F_3 appears as well, then we replace F_2+F_3 by F_4. This is then the last term, unless F_5 appears as well, in which case we replace F_4+F_5 by F_6. If necessary, we proceed in this manner and find that in the canonical representation of $a(N)$ the last Fibonacci term has an even subscript.

In the same way we conclude that in the canonical representation of $b(N)$ the last Fibonacci term has an odd subscript.

It follows that no $a(N)$ can also be a $b(N)$, and that any positive integer is either an $a(N)$ or a $b(N)$. Any number whose canonical representation has a last summand with an odd (or an even) subscript is an $a(N)$ (or a $b(N)$) for some N.

In many applications it is convenient to write

$$F_{k_1}+\ldots+F_{k_r}$$

more concisely as a sequence of 1 and 0 digits, 1 appearing in the k_1th, ..., k_rth position, and 0 in all the others. For instance, the canonical representation of $25 = 21 + 3 + 1$ is then 10001010, and a possible non-canonical representation is $13 + 8 + 3 + 1$, that is 1101010.

X

Search and games

1. Fibonacci numbers turn up in numerical analysis, when it is required to find the minimum or the maximum of a unimodal function within a given interval.

Suppose we want to find the minimum of a convex function in the interval $[AB]$, by subdividing this interval into successively smaller portions, to find that portion within which the minimum will lie. We decide that we shall perform n evaluations of the function, and we denote the length of the interval $[AB]$ by G_{n+1}. Our procedure is as follows:

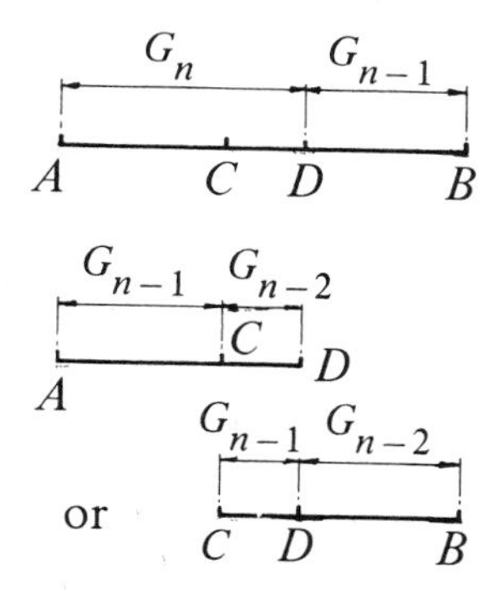

Fig. VI.

Find point C and D such that the lengths $[\overline{AC}] = [\overline{DB}] < \frac{1}{2}[\overline{AB}]$ and denote the intervals as indicated in Fig. VI; their lengths are G_n and G_{n-1}.

Compute the values of the function at C and at D say $f(C)$ and $f(D)$. If $f(C) < f(D)$, then the minimum will be in $[AD]$, while if $f(D) > f(C)$, then it will be in $[CB]$.

We concentrate now on one of the latter intervals, whichever appears to be of interest. One portion of these is already marked, that is has length G_{n-1}, and we mark the remainder G_{n-2}. If we have chosen $[AD]$ for further consideration, then we find now a point E such that $[\overline{AE}] = G_{n-2}$; if we have chosen $[CB]$, then we find a point E such that $[\overline{EB}] = G_{n-2}$. This leads to a third evaluation, at E.

Proceeding in this manner, we reach eventually G_3, subdivided into G_1 and G_2, and evaluate the function at the point where the two intervals meet. This will be the nth evaluation.

The question arises: which particular sequence should we choose for defining the subdivisions?

We may choose all ratios G_{i+1}/G_i to equal τ. This procedure has been called the Golden Section Search. Alternatively, we may choose G_{i+1}/G_i to equal F_{i+1}/F_i: the 'Fibonacci Search'.

The efficiency of the searching procedure depends on the reduction of the length of the given interval $[AB]$ to one of the last intervals. For the Golden Section Search we have $\tau^n/\tau = \tau^{n-1}$. For the Fibonacci Search $F_{n+1}/F_2 = F_{n+1}$. For large n this is approximately $\tau^{n+1}/\sqrt{5}$. The reduction in the latter case is larger, because

$$\frac{\tau^{n+1}/\sqrt{5}}{\tau^{n-1}} = \frac{\tau^2}{\sqrt{5}} \sim 1.17.$$

Therefore, from this point of view, the Fibonacci Search is preferable. Kiefer (1953) has shown that it is optimal within a large set of possible strategies.

Bellman and Dreyfus (1962) give an equivalent result, said to be due to 0. Gross and S. Johnson, by showing that if the extremum of a unimodal function on an interval $0 \leqslant x \leqslant L_n$ is to be located within an interval of unit length, by computing not more than n functional values, then L_n must not be larger than (in our notation) the Fibonacci number F_{n+1}.

The Fibonacci Search finishes with F_1 and F_2, that is with two equal intervals. To decide which of these to examine further, we would choose a point near to the centre of the interval with length F_3, where the two final intervals meet.

The initial choice of n depends on the interval reduction which we wish to achieve.

Example. Find the smallest, value of $y = 3x^2 - 2x + 1$ in the interval [1/4, 3/4], after reducing the lengths of the intervals to 0.1 (a five-fold reduction).

$F_5 = 3 + 2$ 0.25 0.45 0.55 0.75
$y(0.45) = 0.7075,\ y(0.55) = 0.8075$

$F_4 = 2 + 1$ 0.25 0.35 0.45 0.55
$y(0.35) = 0.6675,\ y(0.45) = 0.7075$

$F_3 = 1 + 1$ 0.25 0.35 0.45
$y(0.34) = 0.6668.$

Evaluations at 0.55, 0.45, 0.35, 0.34.
The minimum lies, in fact, at 1/3 and equals 0.6667.

2. Fibonacci Nim, a game designed by Whinihan (1963), is played with one pile of counters, according to the following rules:

Two players, A and B, take counters alternately, A starting.
To begin with, A takes any number of counters, but at least one, and not the whole pile.
After the first move by A, either player takes at least one counter, but not more than twice the number taken by the opponent at the immediately preceding move.
The player who takes the last counter wins.

We shall now show that A can win, provided the number of counters is not a Fibonacci number at the start. The winning strategy is as follows:
If possible take the whole pile. Otherwise express the number of counters still in the pile in a Fibonacci representation, say $N = H_1 + H_2 + \ldots + H_k$. Take the smallest number H_i, that is H_k, from the pile.
B cannot use the winning strategy now. There were originally at least two numbers in the representation, because we assumed that we do not start with a Fibonacci number, and we shall see presently that B cannot reduce the number in the pile to another Fibonacci number. So there is at least one number still left in the list, and B cannot take the smallest number still in the list, because of Property 1 proved in Chapter IX. B takes therefore a smaller number, say D, reducing the pile to N', say.
Now it is A's turn again. He constructs his list for N', by omitting H_k from his previous list and expressing $H_{k-1} - D$ as a sum of Fibonacci numbers, thus $H'_{k-1} + H'_k + \ldots + H'_L$.
The next higher Fibonacci number above H'_{k-1} is H_{k-1}, and by Property 2 (in Chapter IX)

$$H'_L \leqslant 2(H_{k-1} - (H_{k-1} - D)) = 2D \ .$$

It follows that A can now once more use the winning strategy, by taking H'_L.
If, contrary to our assumption, the pile consists originally of a Fibonacci number of counters, say F_i, then A can not help giving his opponent the opportunity of applying the winning strategy for his benefit. A cannot take the whole pile, nor can he put B into the losing position by leaving him with a single Fibonacci number, say $F_i - F_{i-1}$, because then B could take the whole file and win outright.
An example.

Initial pile 32.

	A takes	B takes	Remainder
32 = 21 + 8 + 3	3	6(say)	23
23 = 21 + 2	2	4(say)	17
17 = 13 + 3 + 1	1	2(say)	14
14 = 13 + 1	1	2(say	11
11 = 8 + 3	3	2*	6
6 = 5 + 1	1	1*	4
4 = 3 + 1	1	1 or 2	2 or 1
2 or 1	all, wins.		

*If B took more, A could win at the next move.

3. We now describe a second game, this one played with two piles of counters. It was devised by W. A. Wythoff (1907).

Two players, A and B, take counters in turn, any number from one pile, or the same number from both piles. Each player must take at least one counter at his turn, and the player who takes the last counter wins.

Instead of talking of piles of counters, we can describe the game as being played by moving on a plane between its points with non-negative integer coordinates.

Start at some given point. The players move alternately through any distance, but only south, west, or southwest. The player who reached the point (0,0) first wins. This is the form in which Isaacs (1958) re-invented the game. Kenyon (1967) pointed out the equivalence of the two games, which were, apparently, played much earlier in China.

We construct the winning strategy by working recursively from (0,0). This point can be reached in one move from its diagonal, its column, or its row (see Fig. VII(a)).

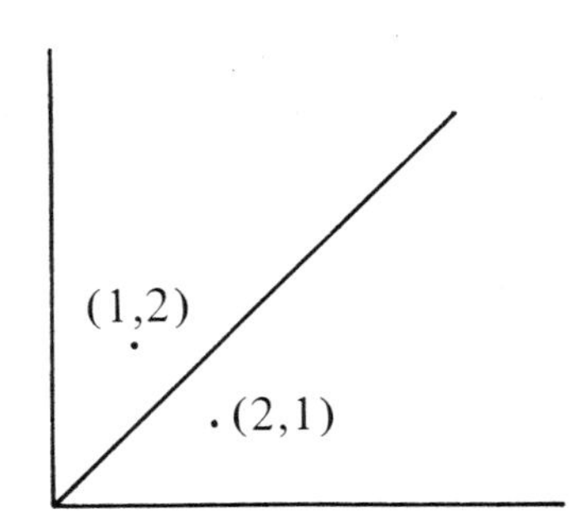

Fig. VIIa.

Therefore the first player, A, must not move to any point on any of these lines. The 'safe' points for A to move to, nearest to (0,0), are therefore (1,2) and (2,1).

These safe points, as well as all the others still to be determined, have the following two properties:

(i) the opponent cannot reach (0,0) or any other safe point in one step from there, and

(ii) wherever B moves to, A can then either reach (0,0) straightaway, or he can move to a safe point.

It follows that further safe points have the property that they have no vertical, horizontal, or diagonal connections to any of those already found to be safe, to ensure property (i). Moreover, a safe point must be such that all points south, west or southwest of it lie on some line emanating vertically, horizontally, or diagonally, from a point known to be safe. This ensures property (ii).

In this way we obtain Fig. VII(b)

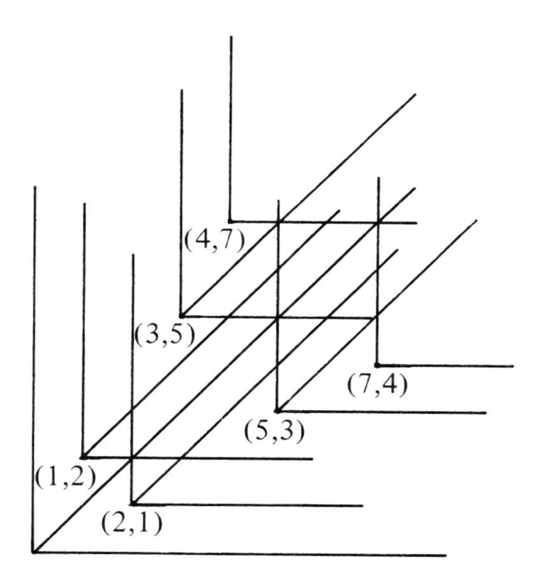

Fig. VIIb.

The difference between the two coordinates of any point increases by 1 as we construct the next point. There is a safe point on every horizontal and on every vertical. The only safe point on the main (central) diagonal is the point (0,0).

Numerically, we can generate the safe points as follows, describing now those points above the central diagonal. Each safe point has a corresponding point symmetrically below that diagonal.

Start with (1,2). Increase the difference between the two coordinates by unity, and choose for the next abscissa the smallest positive integer which has not appeared yet, either as an abscissa or as an ordinate. Thus we obtain

(1,2) (3,5) (4,7) (6,10) 8,13) ...

Let us call an abscissa an a-number, and a coordinate a b-number. The pairs (a,b) are uniquely determined by the following three characteristics (see Carlitz *et al.*, 1972a):

(1) The a-numbers form a monotone sequence, and so do the b-numbers.
(2) The differences between a b-number and the a-number of the same point form the sequence 1,2,3,...
(3) Each positive integer appears either as an a-number, or as a b-number, but none is an a-number and also a b-number.

It will be seen that these are precisely the three characteristics (C_i) we mentioned in Chapter IX, Section 4). We show now that they characterise also the pairs $[n\tau]$ and $[n\tau^2]$, where $[x]$ means the largest integer not exceeding x.

To show this, we prove first a few facts about $[x]$ functions.

Denote $n\tau - [n\tau]$ by f_n. Then

(a) $$[n\tau^2] = [n + n\tau] = n + [n\tau].$$

(b) $$[[n\tau]\tau] = [(n\tau - f_n)\tau] = [n + n\tau - f_n\tau]$$
$$= [n + n\tau - f_n + (1-\tau)f_n] = n + [n\tau] - 1 = [n\tau^2] - 1 \ .$$

(c) $$[n\tau^2]\tau = [(n\tau^2 - f_n)\tau] = [n\tau + n\tau^2 - f_n\tau]$$
$$= [n\tau + n\tau^2 - 2f_n + (2-\tau)f_n] = [(n\tau - f_n) + (n\tau^2 - f_n) + (2-\tau)f_n$$
$$= [n\tau] + [n\tau^2] \ .$$

We show now that the three characteristics (C_1), (C_2) and (C_3) of $a(N)$ and $b(N)$ which we have derived in Section 4 of Chapter IX apply also to the pairs $[n\tau]$ and $[n\tau^2]$.

(C_1) is obvious
(C_2) is demonstrated in (a) above
(C_3) To show this, we prove first that for any N, either

$$\left[\left[\frac{N}{\tau}\right]\tau\right] \quad \text{or} \quad \left[\left[\frac{N}{\tau^2}\right]\tau^2\right]$$

equals $N-1$. Assume the contrary, viz. that $\left[\frac{N}{\tau}\right]\tau < N-1$, and also $\left[\frac{N}{\tau^2}\right]\tau^2 < N-1$. (They can not be N or larger than N.) Then

$$\left[\frac{N}{\tau}\right] < (N-1)/\tau \quad \text{and} \quad \left[\frac{N}{\tau^2}\right] < (N-1)/\tau^2 \ .$$

Since $1/\tau + 1/\tau^2 = 1$, this would mean that $\left[\frac{N}{\tau}\right] + \left[\frac{N}{\tau^2}\right] < N-1$.

However, the left-hand side is an integer, and it cannot possibly be as small as $N-2$ or smaller, because it differs from $N/\tau + N/\tau^2$ by the sum of two fractions, which can not add up to 2 or more.

We must still show that the second part of characteristic (C_3) holds; that is that no $[n\tau]$ can be a $[m\tau^2]$, n and m being integers.

Suppose this were possible, and $[n\tau] = [m\tau^2] = s$, an integer. Then $n\tau - 1 < s < n\tau$ and $m\tau^2 - 1 < s < m\tau^2$. It would follow that

$$(n\tau - 1)/\tau + (m\tau^2 - 1)/\tau^2 < s/\tau + s/\tau^2 < n + m \ ,$$

or $n + m - 1 < s < n + m$. But this is impossible, because s is an integer. (This proof appears in Carlitz *et al.*, 1972a).

Now $[\tau] = a(1)$ and $[\tau^2] = b(1)$ ($a(N)$ and $b(N)$ are defined in Chapter IX, section 3). It follows that

$$a(n) = [n\tau] \quad \text{and} \quad b(n) = [n\tau^2] \ .$$

Explicitly, if

$$n = F_{k_1} + \ldots + F_{k_r}$$

then

$$[n\tau] = f_{k_1+1} + \ldots + F_{k_r+1}$$

and

$$[n\tau^2] = F_{k_1+2} + \ldots + F_{k_r+2}$$

The following table of a-numbers and b-numbers exhibits all the properties which we have proved for them. Note, in particular, the fact that the move from $[n\tau^2]$ to $[n\tau]$ is a 'shift', and so is the move from $[n\tau]$ to n. In the canonical representation all a-numbers have the last Fibonacci term with an even subscript, and all b-numbers have the last Fibonacci term with an odd subscript.

n	$n\tau$	$n\tau^2$	$[n\tau]$		$[n\tau^2]$	
1	1.618..	2.618..	1	10	2	100
2	3.236..	5.236..	3	1000	5	10000
3	4.854..	7.854..	4	1010	7	10100
4	6.472..	10.472..	6	10010	10	100100
5	8.090..	13.090..	8	100000	13	1000000
6	9.708..	15.708..	9	100010	15	1000100
7	11.326..	18.326..	11	101000	18	1010000
8	12.944..	20.944..	12	101010	20	1010100
9	14.562..	23.562..	14	1000010	23	10000100
10	16.180..	26.180..	16	1001000	26	10010000

A possible game.
Start with (20,26)

A takes	to establish	B takes	to establish
(4,0)	(16,26)	(0,1)	(16,25)
(2,2)	(14,23)	(3,0)	(11,23)
(0,5)	(11,18)	(2,2)	(9,16)
(0,1)	(9,15)	(0,2)	(9,13)
(3,3)	(6,10)	(0,8)	(6, 2)
(5,0)	(1, 2)		
and wins at his next turn			

We can also make a statement about the fractional parts of $n\tau$. They can be considered to be the distances from a point O in a circle with circumference 1, if the lengths $n\tau$ are marked on the circumference, starting from O. Denote the fractions by $<n\tau> = n\tau - [n\tau]$.

Consider a straight line of length 1, representing the circumference straightened out. We draw on it the point $<\tau>$, that is at a distance of $-\sigma$ from the left-hand endpoint, since $\tau - 1 = -\sigma$ (see Fig. VII(c) (i)). The point divides the line of length 1 in the Golden Section.

Now draw the point $<2\tau>$. It falls within the larger of the new intervals, dividing it again in the Golden Section. We have now three intervals, two of them being of equal length (see VII(c) (ii)).

The point $<3\tau>$ divides one of the larger intervals, again in the same ratio (Fig. VII(c) (iii)).

We continue in this manner. Fig. VII(c) (iv) gives the example after three more steps.

At every stage, the new point will divide one of the largest intervals just

0.618	0.382	(i)

0.618		0.382	(ii)
0.236	0.382	0.382	

0.618		0.382		(iii)
0.236	0.382	0.382		
0.236	0.382	0.236	0.146	

0.618					0.382			(iv)
0.236		0.382			0.382			
0.236		0.382			0.236		0.146	
0.236		0.236	0.146		0.236		0.146	
0.090	0.146	0.236	0.146		0.236		0.146	
0.090	0.146	0.236	0.146		0.090	0.146	0.146	

0 0.090 0.236 0.472 0.618 0.708 0.834 1.000

Fig. VIIc

constructed. This was conjectured, in a different but equivalent formulation, by H. Steinhaus and proved by Świerczkowski (1958). The proof is simple, but too long to be quoted here.

XI

Hyperbolic functions and Fibonacci numbers

There exists a close analogy between formulae concerning a hyperbolic function with argument $\frac{1}{2}n \ln \tau/\sigma$, and a Fibonacci number F_n or L_n.

By definition

$$\sinh nz = \tfrac{1}{2}(e^{nz} - e^{-nz})$$

and

$$\cosh nz = \tfrac{1}{2}(e^{nz} + e^{-nz}) \ .$$

Let $z = \frac{1}{2} \ln \tau/\sigma$, then

$$\sinh nz = \frac{1}{2}\left[\left(\frac{\tau}{\sigma}\right)^{\frac{1}{2}n} - \left(\frac{\sigma}{\tau}\right)^{\frac{1}{2}n}\right] = \frac{1}{2}\frac{\tau^n - \sigma^n}{(\tau\sigma)^{\frac{1}{2}n}}$$

$$= \frac{F_n\sqrt{5}}{2\mathrm{i}^{\mathrm{n}}}$$

and

$$\cosh nz = \frac{1}{2}\left[\left(\frac{\tau}{\sigma}\right)^{\frac{1}{2}n} + \left(\frac{\sigma}{\tau}\right)^{\frac{1}{2}n}\right] = \frac{1}{2}\frac{\tau^n + \sigma^n}{(\tau\sigma)^{\frac{1}{2}n}}$$

$$= \frac{L_n}{2i^n} \ .$$

It follows, that any relationship between hyperbolic functions is analogous to a relationship between Fibonacci and Lucas numbers. We shall now give some examples.

(1) $$\cosh nz + \sinh nz = e^{nz} = \tau^n/i^n$$

$$\cosh nz - sinh\, nz = e^{-nz} = \sigma^n/i^n$$

is valid when $z = \frac{1}{2}\ln(\tau/\sigma)$.

Hence

$$\frac{L_n}{2i^n} + \frac{\sqrt{5}F_n}{2i^n} = \frac{\tau^n}{i^n}, \quad \text{i.e.} \quad L_n + \sqrt{5}F_n = 2\tau^n .$$

$$\frac{L_n}{2i^n} - \frac{\sqrt{5}F_n}{2i^n} = \frac{\sigma^n}{i^n}, \quad \text{i.e.} \quad L_n - \sqrt{}F_n = 2\sigma^n .$$

(2) $$\cosh^2 nz - \sinh^2 nz = 1 ,$$

hence

$$\frac{L_n^2}{4(-1)^n} - \frac{5F_n^2}{4(-1)^n} = 1 , \quad \text{or} \quad 5F_n^2 - L_n^2 = 4(-1)^{n+1} ,$$

our formula (24).

(3)
$$\begin{aligned}
\sinh(m+n)z + \sinh(m-n)z &= 2\sinh mz \cosh nz \\
\sinh(m+n)z - \sinh(m-n)z &= 2\cosh mz \sinh nz \\
\cosh(m+n)z + \cosh(m-n)z &= 2\cosh mz \cosh nz \\
\cosh(m+n)z - \cosh(m-n)z &= 2\sinh mz \sinh nz
\end{aligned}$$

These are, respectively, analogous to

$$\begin{aligned}
F_{m+n} + F_{m-n}(-1)^n &= F_m L_n, \text{ formula (15a)} \\
F_{m+n} - F_{m-n}(-1)^n &= L_m F_n, \text{ formula (15b)} \\
L_{m+n} + L_{m-n}(-1)^n &= L_m L_n, \text{ formula (17a)} \\
L_{m+n} - L_{m-n}(-1)^n &= 5F_m F_n, \text{ formula (17b)}
\end{aligned}$$

(4)
$$\begin{aligned}
\sinh 2nz &= 2\sinh nz \,.\, \cosh nz \\
\cosh 2nz &= \cosh^2 nz + \sinh^2 nz
\end{aligned}$$

These correspond to

$$F_{2n} = F_n L_n \qquad \text{formula (13)}$$

$$2L_{2n} = L_n^2 + 5F_n^2 \qquad \text{formula (22)}$$

(5)
$$\sinh nz = \text{n}\cosh^{n-1} z \sinh z + \binom{n}{3} \cosh^{n-2} z \sinh^3 z + \ldots .$$

and

$$\cosh nz = \cosh^n z + \binom{n}{2} \cosh^{n-3} z \sinh^2 z + \binom{n}{4} \cosh^{n-4} z \sinh^4 z + \ldots$$

correspond to

$$\frac{F_n\sqrt{5}}{2i^n} = \frac{n}{(2i)^{n-1}} \frac{1}{2i} \sqrt{5} + \binom{n}{3} \frac{1}{(2i)^{n-3}} \frac{1}{(2i)^3} (\sqrt{5})^3 + \ldots$$

and

$$\frac{L_n}{2i^n} = \frac{1}{(2i)^n} + \binom{n}{2} \frac{1}{(2\text{i})^{\text{n}-2}} \frac{1}{(2\text{i})^2} (\sqrt{5})^2 + \ldots$$

that is

$$F_n = \frac{1}{2^{\text{n}-1}} \left[\binom{n}{1} + \binom{n}{3} 5 + \binom{n}{5} 5^2 + \ldots \right], \qquad \text{formula (91)}$$

and

$$L_n = \frac{1}{2^{n-1}} \left[1 + \binom{n}{2} 5 + \binom{n}{4} 5^2 + \ldots \right], \qquad \text{formula (92).}$$

(6) Yet another formula with hyperbolic functions, which is valid when $z = \frac{1}{2} \ln \tau/\sigma$ is

$$\sinh(\text{n}+1) z \sinh(n-1) z - (\sinh)^2 = 5(-1)^n ,$$

which is analogous to

$$F_{n+1}F_{n-1} - F_n^2 = (-1)^n , \qquad \text{our formula (29).}$$

XII

Meta-Fibonacci sequences (a letter from B. W. Conolly)

As you are writing a book about Fibonacci sequences and related topics, I thought you might like to consider mentioning a family of fascinating and apparently puzzling sequences about which I am currently reading more and more.

These sequences, like those of Fibonacci, are generated by linear combinations of previous sequence values. However, the Fibonacci (F_n) is generated by $F_n = F_{n-1} + F_{n-2}$ and it is always the immediately preceding values F_{n-1} and F_{n-2} that determine F_n. By contrast, the sequences of which I write, say (N_n), are such that N_n is determined by the sum of two preceding values, just like F_n, but which preceding values depend on n and prior values of N_n. I am going to discuss briefly four such sequences defined for positive integer arguments as follows.

$$\begin{aligned} H(1) &= H(2) = 1 , \\ H(n) &= H(n - H(n-1)) + H(n - H(n-2)) , \quad (n > 2) ; \end{aligned} \tag{1}$$

$$\begin{aligned} F(1) &= 0 \text{ or } 1 , \ F(2) = 1 , \\ F(n) &= F(n - F(n-1)) + F(n - 1 - F(n-2)) , \quad (n > 2); \end{aligned} \tag{2}$$

Note that this $F(n)$ is not your Fibonacci F_n.

$$\begin{aligned} C(1) &= C(2) = 1 , \\ C(n) &= C(n - C(n-1)) + C(C(n-1)) , \quad n > 2, \end{aligned} \tag{3}$$

$$\begin{aligned} K(1) &= K(2) = 1 , \\ K(n) &= K(K(n-1)) + K(K(n-2)) , \quad n > 2 , \end{aligned} \tag{4}$$

For ease of notation I am writing as arguments what are usually written as subscripts.

$(H(n))$ is the sequence that first aroused my interest and remains the one about which, apparently, least is known. (2) and (3) are obviously close relatives and I know at least how to express them explicitly as functions of n, i.e. to solve the recurrences. (4) looks as odd as the others but is the simplest. The minute fragment of information I give in this letter seems just to scratch the surface of an intriguing puzzle and I imagine that it will tempt the more adventurous among your readers to investigate these curiosities and generalizations of them more deeply for themselves.

1. It will be useful to start the discussion with some values of the sequences. I give these in the Table 1 for $n = 1$ to $n = 261$. There is a reason for this choice. It turns out to be important to examine the values over successive 'octaves', by which I mean over ranges of n from 2^m *to* $2^{m+1} - 1$. Powers of 2 do indeed play a central role in the little I know about the sequences.

A first glance at the Table 1 reveals that $K(n) = 2$ for $n > 2$. I might not have repeated it, but it is useful to see it so as not to forget its behaviour. In fact $K(n)$ is not as simple as it looks. With other than the initial values $K(1) = K(2) = 1$ it is soon found that the sequence runs into surprises. With, for example, $K(1) = 1$, $K(2) = 2$ we get $K(3) = 3$, $k(4) = 5$ and thereafter $K(n)$ has to be ∞ to make any sense. The sequence is indeed very finely balanced on a knife edge and the sort of behaviour I have just described is strongly reminiscent of the chaos in the modern technical sense that is quite fashionable.

Table 1 shows that apparently $(F(n))$ and $(C(n))$ are monotonic non-decreasing. $C(n)$ shows a tendency to pull away from $F(n)$ for quite long periods, but they seem to come together again quite regularly. Of the two, $F(n)$ appears the more sluggish. As noted above, the $(F(n))$ sequence is the same for $n > 2$ when $F(2) = 1$ and $F(1)$ is either 0 or 1. I shall show presently that it is more convenient when representing the solution of (2) to use $F(1) = 0$. This is not the case with $(C(n))$ which shares some of the instability mentioned above in the case of $(K(n))$, for, using the boundary values $C(1) = 0$, $C(2) = 1$ results in $C(n) = 1$ for $n \geqslant 3$, whereas with $C(1) = 1$, $C(2) = 2$ we find $C(n) = n$ for $n \geqslant 3$. These anomalies show that the sequences are more awkward than Fibonacci sequences, an observation exacerbated by the apparent lack of a general theory.

$(H(n))$ seems to behave in an 'on the average' sense roughly like $(C(n))$ and $(F(n))$ in that there seems to be a common trend, but its fine structure is quite ragged. For a time it behaves with exemplary docility only to tear away from the others by making a big leap which is later compensated by leaps in the opposite direction (see, for example, $n = 191, 192, 193$). It is certainly not monotonic in the usual sense. Moreover, some values do not appear at all — 7, 15 and 18 are early examples.

2. I should explain next how I came to be interested in these sequences. I encountered $(H(n))$ for the first time in an article by Richard Guy [1]. Guy seems to be the one who coined the description 'meta-Fibonacci' for the sequences. He had mentioned $(H(n))$ also briefly in [2], but the inventor seems to have been Hofstadter [3] who, in a remarkable book, introduces the sequence in a chapter on recursive structures in a number of interesting applications including music, art and language.

Hofstadter describes $(H(n))$ as a 'chaotic sequence' and says that it leads to 'a small mystery . . . the further out you go' (in the sequence) '. . . the less sense it seems to make . . . what is of interest is whether there is another way of characterizing this sequence and with luck, a nonrecursive way.' Guy [1] describes calculations reported by correspondents which do not clarify the issue of solving the recurrence.

Table 1 — Values of some 'meta-Fibonacci' sequences

n	$H(n)$	$F(n)$	$C(n)$	$K(n)$	n	$H(n)$	$F(n)$	$C(n)$	$K(n)$
1	1	1	1	1	46	24	24	27	2
2	1	1	1	1	47	24	24	27	2
3	2	2	2	2	48	32	25	27	2
4	3	2	2	2	49	24	26	28	2
5	3	3	3	2	50	25	26	29	2
6	4	4	4	2	51	30	27	29	2
7	5	4	4	2	52	28	28	30	2
8	5	4	4	2	53	26	28	30	2
9	6	5	5	2	54	30	28	30	2
10	6	6	6	2	55	30	29	31	2
11	6	6	7	2	56	28	30	31	2
12	8	7	7	2	57	32	30	31	2
13	8	8	8	2	58	30	31	31	2
14	8	8	8	2	59	32	32	32	2
15	10	8	8	2	60	32	32	32	2
16	9	8	8	2	61	32	32	32	2
17	10	9	9	2	62	32	32	32	2
18	11	10	10	2	63	40	32	32	2
19	11	10	11	2	64	33	32	32	2
20	12	11	12	2	65	31	33	33	2
21	12	12	12	2	66	38	34	34	2
22	12	12	13	2	67	35	34	35	2
23	12	12	14	2	68	33	35	36	2
24	16	13	14	2	69	39	36	37	2
25	14	14	15	2	70	40	36	38	2
26	14	14	15	2	71	37	36	38	2
27	16	15	15	2	72	38	37	39	2
28	16	16	16	2	73	40	38	40	2
29	16	16	16	2	74	39	38	41	2
30	16	16	16	2	75	40	39	42	2
31	20	16	16	2	76	39	40	42	2
32	17	16	16	2	77	42	40	43	2
33	17	17	17	2	78	40	40	44	2
34	20	18	18	2	79	41	40	45	2
35	21	18	19	2	80	43	41	45	2
36	19	19	20	2	81	44	42	46	2
37	20	20	21	2	82	43	42	47	2
38	22	20	21	2	83	43	43	47	2
39	21	20	22	2	84	46	44	48	2
40	22	21	23	2	85	44	44	48	2
41	23	22	24	2	86	45	44	48	2
42	23	22	24	2	87	47	45	49	2
43	24	23	25	2	88	47	46	50	2
44	24	24	26	2	89	46	46	51	2
45	24	24	26	2	90	48	47	51	2

Table 1 — *Continued*

n	$H(n)$	$F(n)$	$C(n)$	$K(n)$	n	$H(n)$	$F(n)$	$C(n)$	$K(n)$
91	48	48	52	2	136	61	69	71	2
92	48	48	53	2	137	71	70	72	2
93	48	48	53	2	138	77	70	73	2
94	48	48	54	2	139	65	71	74	2
95	48	48	54	2	140	80	72	75	2
96	64	49	54	2	141	71	72	76	2
97	41	50	55	2	142	69	72	76	2
98	52	50	56	2	143	77	72	77	2
99	54	51	56	2	144	75	73	78	2
100	56	52	57	2	145	73	74	79	2
101	48	52	57	2	146	77	74	80	2
102	54	52	57	2	147	79	75	80	2
103	54	53	58	2	148	76	76	81	2
104	50	54	58	2	149	80	76	82	2
105	60	54	58	2	150	79	76	83	2
106	52	55	58	2	151	75	77	83	2
107	54	56	59	2	152	82	78	84	2
108	58	56	60	2	153	77	78	85	2
109	60	56	60	2	154	80	79	85	2
110	53	56	61	2	155	80	80	86	2
111	60	57	61	2	156	78	80	86	2
112	60	58	61	2	157	83	80	86	2
113	52	58	62	2	158	83	80	87	2
114	62	59	62	2	159	78	80	88	2
115	66	60	62	2	160	85	81	89	2
116	55	60	62	2	161	82	82	90	2
117	62	60	63	2	162	85	82	90	2
118	68	61	63	2	163	84	83	91	2
119	62	62	63	2	164	84	84	92	2
120	58	62	63	2	165	88	84	93	2
121	72	63	63	2	166	83	84	93	2
122	58	64	64	2	167	87	85	94	2
123	61	64	64	2	168	88	86	95	2
124	78	64	64	2	169	87	86	95	2
125	57	64	64	2	170	86	87	96	2
126	71	64	64	2	171	90	88	96	2
127	68	64	64	2	172	88	88	96	2
128	64	64	64	2	173	87	88	97	2
129	63	65	65	2	174	92	88	98	2
130	73	66	66	2	175	90	89	99	2
131	63	66	67	2	176	91	90	99	2
132	71	67	68	2	177	92	90	100	2
133	72	68	69	2	178	92	91	101	2
134	72	68	70	2	179	94	92	101	2
135	80	68	71	2	180	92	92	102	2

Table 1 — *Continued*

n	$H(n)$	$F(n)$	$C(n)$	$K(n)$
181	93	92	102	2
182	94	93	102	2
183	94	94	103	2
184	96	94	104	2
185	94	95	104	2
186	96	96	105	2
187	96	96	105	2
188	96	96	105	2
189	96	96	106	2
190	96	96	106	2
191	96	96	106	2
192	128	97	106	2
193	72	98	107	2
194	96	98	108	2
195	115	99	109	2
196	100	100	109	2
197	84	100	110	2
198	114	100	111	2
199	110	101	111	2
200	93	102	112	2
201	106	102	112	2
202	124	103	112	2
203	82	104	113	2
204	101	104	114	2
205	111	104	114	2
206	108	104	115	2
207	118	105	115	2
208	104	106	115	2
209	108	106	116	2
210	106	107	116	2
211	114	108	116	2
212	104	108	116	2
213	114	108	117	2
214	109	109	118	2
215	100	110	118	2
216	109	110	119	2
217	120	111	119	2
218	112	112	119	2
219	108	112	120	2
220	118	112	120	2
221	106	112	120	2
222	105	112	120	2
223	130	113	121	2
224	110	114	121	2
225	114	114	121	2
226	115	115	121	2
227	112	116	121	2
228	107	116	122	2
229	120	116	123	2
230	114	117	123	2
231	122	118	124	2
232	121	118	124	2
233	120	119	124	2
234	114	120	125	2
235	138	120	125	2
236	110	120	125	2
237	122	120	125	2
238	119	121	126	2
239	120	122	126	2
240	130	122	126	2
241	132	123	126	2
242	113	124	126	2
243	133	124	127	2
244	123	124	127	2
245	118	125	127	2
246	125	126	127	2
247	121	126	127	2
248	129	127	127	2
249	122	128	128	2
250	136	128	128	2
251	129	128	128	2
252	116	128	128	2
253	149	128	128	2
254	137	128	128	2
255	120	128	128	2
256	123	128	128	2
257	143	129	129	2
258	146	130	130	2
259	107	130	131	2
260	139	131	132	2
261	138	132	133	2

This article inpsired the recurrence (2) which I felt to have greater symmetry than (1), and when I found that I could solve it I wrote in August 1986 to Guy. He replied later in the same month expressing interest and drawing attention to the recurrence (3) which he attributed to John Conway, who at that time was unable to furnish a solution. The latter also referred to an as yet unpublished article by S. Golomb to be entitled 'Sequences satisfying 'strange' recursions' and still, as far as I know, unpublished, and indicated nine properties of $C(n)$ claimed by Golomb to be provable, including the statement: 'Formulas can be given (for $C(n)$) by relating n to the 'nearest' power of 2.' However, no formula was given, nor any indication that the structure of $(C(n))$ had been identified. Challenged by this I experimented and found that I could identify exactly the structure of (3) and communicated my findings to Guy in October, 1986. Guy replied that he would send my observations to Conway, and that terminated the correspondence until 29 May, 1988, when, puzzled by Guy's failure to mention our correspondence in a review paper of unsolved problems [4], I wrote to him. The answer indicated that he had mislaid my work. He suggested that I write to Golomb for a preprint of his paper, and that letter received no reply.

An interesting postscript is, however, furnished by an article in the *New York Times* of August 30, 1988. This reported that Conway had expounded his sequence at a seminar in July, 1988, at the Bell Laboratories, New Jersey, and had offered a prize of \$10 000, though he meant \$1000, essentially for a solution of the recurrence. A solution was then furnished by Colin Mallows of the Bell Labs and the affair of the mistaken prize offer was revealed. An amusing anecdote, particularly when I felt that I could have given the answer two years earlier! I have not seen the Mallows solution.

3. To conclude I shall discuss briefly the sequences $(F(n))$ and $(C(n))$. First $(F(n))$. Let $n \geqslant 2$ and let m be the largest integer such that

$$n = 2^m + k \ , \qquad (m \geqslant 1, 0 \leqslant k \leqslant 2^m - 1) \ . \tag{5}$$

Then, with $F(1) = 0$, which yields the same sequence from $n = 3$ onwards by calculation from the recurrence as with $F(1) = 1$, the solution of (2), valid for $n \geqslant 2$, is

$$F(n) = 2^{m-1} + F(k+1) \ , \qquad n \geqslant 2 \ . \tag{6}$$

In effect, (5) represents n in the binary scale, and similarly (6) represents $F(n)$ in the same scale. Indeed, the binary scale seems 'natural' for these recurrences.

It is readily checked that (6) gives the corresponding values in the Table. The proof that it holds generally for positive n is by induction on n.

First we show that $n > F(n)$ in some octave $(2^m, 2^{m+1} - 1)$ follows if the inequality holds over the range $(0, 2^m - 1)$. Thus, representing n by (5), we have

$$\begin{aligned} n - F(n) &= 2^m + k - 2^{m-1} - F(k+1) = \\ &\quad 2^{m-1} + k - F(k+1) > 2^{m-1} + k_{\min} - F(k_{\max} + 1) \\ &\quad = 2^{m-1} - F(2^m) = 0, \text{ by (6)}, \end{aligned}$$

as required.

Now suppose that (6) holds for $n = 2, 3, \ldots, N$. N defines a corresponding M and K such that

$$N = 2^M + K \ .$$

Then,

$$N+1-F(N) = 2^M + K + 1 - 2^{M-1} - F(K+1) = 2^{M-1} + K + 1 - F(K+1)$$
$$N - F(N-1) = 2^M + K - 2^{M-1} - F(K) = 2^{M-1} + K - F(K) \ .$$

Since $K+1 > F(K+1)$ and $K > F(K)$, we get from (6),

$$F(N+1-F(N)) = 2^{M-2} + F(K+2-F(K+1))$$
$$F(N-F(N-1)) = 2^{M-2} + F(K+1-F(K)) \ ,$$

and so

$$F(N+1-F(N)) + F(N-F(N-1)) =$$
$$2^{M-1} + F(K+2-F(K+1)) + (F(K+1-F(K)) =$$
$$2^{M-1} + F(K+2) = F(N+1)$$

after using the defining difference equation (2) and (6). This completes the induction.

In passing it is worth making the observation that Hoftstadter's mechanism (1) admits the particular solution $H(n) = 2^{m-1} + H(k+q)$, where q is some integer. That is to say, if one substitutes this form into the recurrence it appears to be satisfied. However, comparison of the tabulated values indicates that the integer q must depend on n. There is the additional problem that $H(n) - 2^{m-1}$ sometimes assumes values which $(H(n))$ omits!

Let us turn now to the sequence $(C(n))$ defined by (3). This also has the particular solution $C(n) = n$ if the boudnary values are $C(1) = 1$, $C(2) = 2$. In the case put forward by Conway the boundary values are $C(1) = 1$ and $C(2) = 1$. This leads to the much more interesting sequence illustrated by the table. Scrutiny of the values discloses immediately a regular pattern and it is somewhat surprising that the inventor of the sequence, a brilliant and original mathematician, did not spot this for himself.

What I shall say now is descriptive, but can be proved inductively. The pattern is well defined in 'octaves'. The mth octave starts at $n = 2^m$ and terminates at $n = 2^{m+1} - 1$. The initial value in the mth octave is $C(2^m) = 2^{m-1}$. Knowledge of this and the pattern enables $C(n)$ to be constructed for any n within the current octave.

To describe the pattern we need the concept of a 'run-up' of length $k(\geqslant 1)$. In this context, such a run-up starting at an integer a and of length k is the sequence of k successive integers $a, a+1, a+2, \ldots, a+k-1$. An octave of $(C(n))$ consists of sequences of runs-up. The initial member of a given run-up is equal to the last member of the previous run-up.

I shall now describe the $m = 6$th octave in detail, that is $C(n)$ for $n = 64$ to 127. The values in the Table are shown in Fig. 1 in such a way as to illustrate the pattern.

It will be seen that the evolution of $(C(n))$ is a sequence of runs-up. First there is a run-up of 7, and this is followed by 5 runs-up of lengths 5, 4, 3, 2, 1. I have denoted this last sequence by S_5, but its length in terms is $1+2+3+4+5 = \binom{6}{2}$. I shall use the S_k notation, and developments from it, in these two senses, namely to denote a sequence of runs-up of lengths $k, k-1 \ldots, 2, 1$, in that order, accounting for $\binom{k+1}{2}$ terms of $(C(n))$. S_5 is followed by S_4, S_3, S_2, S_1: again I use notation T_k to indicate such a sequence of sequences, namely $T_k = S_k + S_{k-1} + \ldots + S_2 + S_1$, in that order, and also the number of terms accounted for is $\binom{k+2}{3}$.

The succession of sequences to complete the cycle is shown in Fig. 1. I am using in a similar sense

$$U_k = T_k + T_{k-1} + \ldots + T_2 + T_1 = \binom{k+3}{4} \text{ (numerically)} ,$$

$$V_k = U_k + U_{k-1} + \ldots . + U_2 + U_1 = \binom{k+4}{5} ,$$

both to denote successions and the corresponding numbers of terms.

Fig. 1 can be collapsed into a more succinct form:

$$\begin{array}{lll} m = 6 & & \\ \quad 7 & & \\ \quad T_5 & & \text{This is to be read from left to right} \\ \quad U_3 & & \text{and from top to bottom.} \\ 2^{\circ} & U_2 & \\ 2 & \underline{U_1\ U_1} & \end{array}$$

The powers of 2 indicate the number of recurrences of the corresponding U-sequences.

In Fig. 2 I show the cases $m = 7, 8, 9, 10$ in similar form.

In explanation of Fig. 2 I must say that, to save space, the convention is used of displacing triangular blocks vertically. The additional convention is adopted of reading from left to right and top to bottom down to the first horizontal line as if this were the bottom of the page, and then 'turning the page over' and so on repeatedly, each time down to the next horizontal line, or page bottom, and so on.

There are various checks and balances that allow the diagram for any m to be constructed automatically. One of these is the theoretically verifiable fact that after

n	$C(n)$	Length of run-up	S	T	U	V
64–70	32, 33, . . . , 38	7	$= 6+1$	$35 = T_5 = \binom{7}{3}$		
71–75	38, 39, . . . , 42	5	$15 = S_5 = \binom{6}{2}$			
76–79	42, 43, . . . , 45	4				
80–82	45, 46, 47	3				
83–84	47, 48	2				
85	48	1				
86–89	48, 49, . . . , 51	4	$10 = S_4 = \binom{5}{2}$			
90–92	51, 52, 53	3				
93–94	53, 54	2				
95	54	1				
96–98	54, 55, 56	3	$6 = S_3 = \binom{4}{2}$			
99–100	56, 57	2				
101	57	1				
102–103	57, 58	2	$3 = S_2 = \binom{3}{2}$			
104	58	1				
105	58	1	$1 = S_1 = \binom{2}{2}$			
106–108	58, 59, 60	3	$6 = S_3 = \binom{4}{2}$	$10 = T_3 = \binom{5}{3}$	$U_3 = \binom{6}{4}$	$V_2 = \binom{6}{5}$
109–110	60, 61	2				
111	61	1				
112–113	61, 62	2	$3 = S_2 = \binom{3}{2}$			
114	62	1				
115	62	1	$1 = S_1 = \binom{2}{2}$			
116–117	62, 63	2	$3 = S_2 = \binom{3}{2}$	$4 = T_2 = \binom{4}{3}$		
118	63	1				
119	63	1	$1 = S_1 = \binom{2}{2}$			
120	63	1	$1 = S_1 = \binom{2}{2}$	$1 = T_1 = \binom{3}{3}$		
121–122	63, 64	2	$3 = S_2 = \binom{3}{2}$	$4 = T_2 = \binom{4}{3}$	$U_2 = \binom{5}{4}$	
123	64	1				
124	64	1	$1 = S_1 = \binom{2}{2}$			
125	64	1	$1 = S_1 = \binom{2}{2}$	$1 = T_1 = \binom{3}{3}$		
126	64	1	$1 = S_1 = T_1 = U_1 = \binom{4}{4}$			$V_1 = \binom{5}{5}$
127	64	1	$1 = S_1 = T_1 = U_1$			

Fig. 1 — Illustration of pattern for 6th Octave of $(C(n))$.

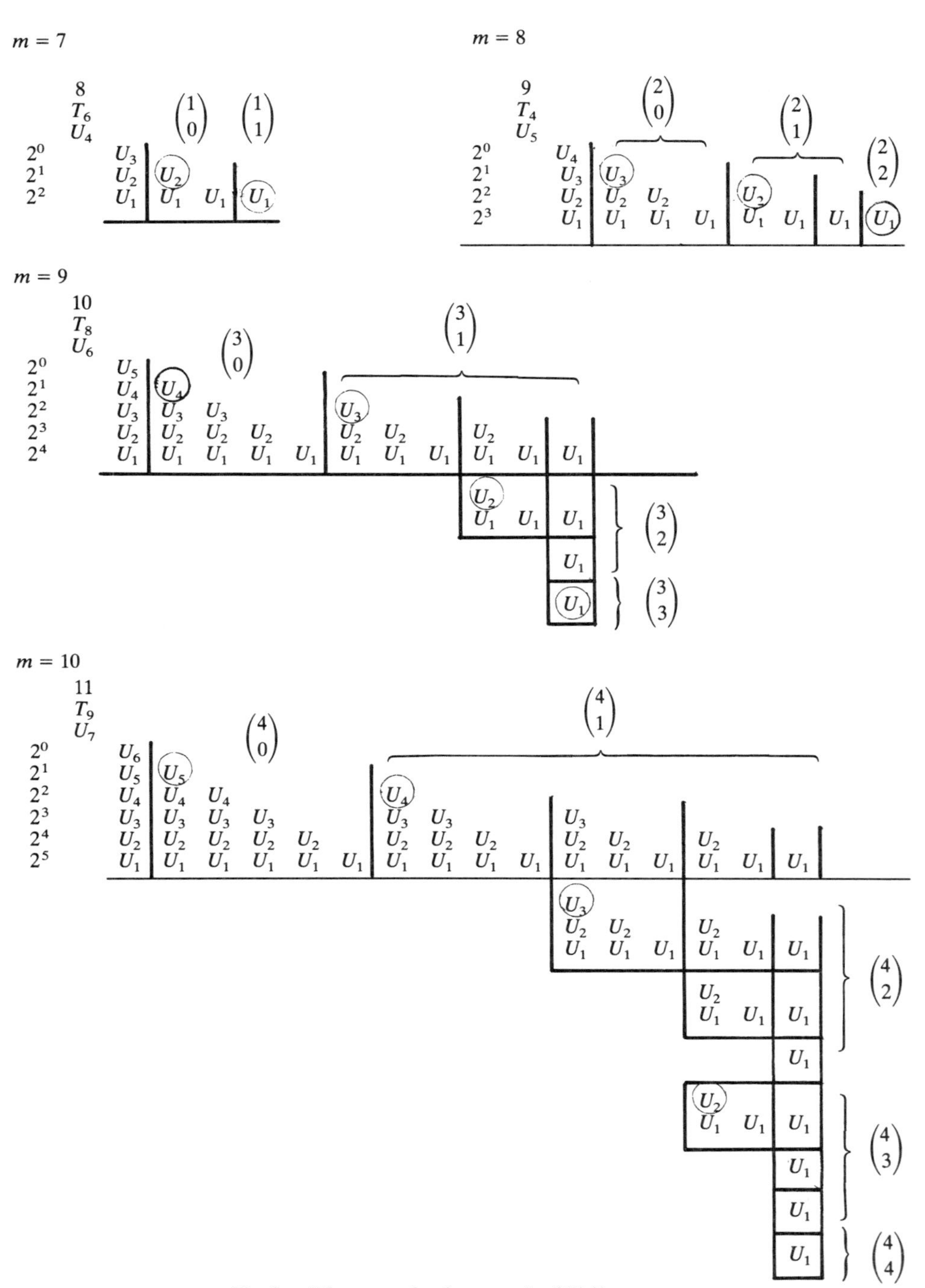

Fig. 2 — Diagrams of mth octave for $(C(n))$ sequence.

the initial sequences of lengths $m+1$, T_{m-1}, U_{m-3}, U_{m-k} for $k = 4$ to $m-1$ occurs exactly 2^{k-4} times each in the whole diagram. To account for all the terms this must mean that

$$m+1+T_{m-1}+U_{m-3}+\sum_{k=4}^{m-1} 2^{k-4}U_{m-k} = 2^m ,$$

and the reader can be invited to prove this identity for himself. It is convenient, as shown, to indicate the powers of 2 at the left of the diagram.

In the diagram the last occurrence of each U_{m-k} from $k=5$ up to $m-1$ is indicated by circling the entry. Each such entry may be thought of as initiating a block consisting of successive triangular arrays of reducing dimension down to U_1. The number of triangular arrays in the block initiated by the last U_{m-k} is $\binom{m-6}{k-5}$. These properties can be incorporated in an algorithm for the automatic construction of the diagram for arbitrary m.

This discussion seems enough to show that $(C(n))$ defined by (3) is as completely understood as $(F(n))$, even though its explanation is more ponderous. It is interesting to note that the solutions given satisfy the initial conditions in both cases (if we take $F(1)=1$). It is also instructive to recall how productive a computational approach has been, and can be in many puzzling problems. Computation reveals patterns leading to conjecture solutions which can be tested theoretically. This approach has not so far been successful for $(H(n))$, but, if this does not admit a solution, and there seems no reason why it should not have a solution, then ultimately, like any code, it will be cracked.

To conclude, here is a problem taken from [5], and proposed by David Newman, Beer Sheva, Israel. Fix a positive integer k. Define $f(n)$ on positive integers by $f(n) = 1$ for $n \leqslant k+1$ and $f(n) = f(f(n-1))+f(n-f(n-1))$ for $n > k+1$. Define the sequence (F_m) by $F_m = 1$ for $m < k$ and $F_m = F_{m-1}+F_{m-k}$ for $m > k$. Prove that

(a) $f_n - f_{n-1} = 0$ or 1 for all n and that f_n is unbounded;

(b) $f(F_{m+k}) = F_m$ for $m \geqslant 1$.

This is a fitting synthesis of the meta-Fibonacci and Fibonacci families with which to end this letter.

References

[1] Guy, Richard (1986) Some suspiciously simple sequences. *Am. Math. Monthly*, **93**, 186–190.
[2] Guy, Richard (1981) *Unsolved problems in number theory*. Springer.
[3] Hofstadter, D. (1979) *Gödel, Escher, Bach:an eternal golden braid*. The Harvester Press, Sussex.

[4] Guy, Richard (1987) Monthly unsolved problems 1969–1987. *Am. Math. Monthly*, **94**, 961–970.
[5] Problem E3274 (1988) *Am. Math. Monthly*, **95**, 555.

XIII

The Golden Section in the plane

1. Divide an interval of length $m+n$ $(m>n)$ into two portions, of lengths m and n respectively, in such a way that

$$(m+n)/m = m/n \ .$$

Then $(m/n)^2 = 1 + (m/n)$, that is $(m/n) = \frac{1}{2}(1 \pm \sqrt{5})$, i.e. τ or σ.

Euclid called τ the 'extreme and mean ratio', while Luca Pacioli (1509) called it the Divine Proportion. At present, it is more usually called the Golden Section. Its value, to ten decimals, is given in Chapter IV, Section 1.

It has been claimed — and it has been contested — that the ratio τ:1 of the sides of a 'Golden Rectangle' gives to the latter a particularly pleasant shape. For instance, the height (with the pediment) and the width of the Parthenon in Athens (see cover) are supposed to fit well into a Golden Rectangle. J. Gordon (1934) has detected a Golden Section in Constable's *The cornfield*, in Rembrandt's *Portrait of a lady*, and in Titian's *Venus and Adonis*, all in the National Gallery in London.

There have also been a number of patently cranky claims. We do not intend to take part in this controversy; we have merely mentioned a few examples to justify the adjective 'golden'.

We have already mentioned (after formula (59)), that the successive powers of τ form a generalized Fibonacci sequence. We quote here a few consequences of (8)

$$\tau^2 = \tau+1, \quad \tau^3 = 2\tau+1, \quad \tau^4 = 3\tau+2$$
$$\tau^{-1} = \tau-1, \quad \tau^{-2} = -\tau+2, \quad \tau^{-3} = 2\tau-3 \ .$$

2. It is very simple to construct with straight edge and compass an interval of length τ, when a length of 1 is given. In a unit square $ABCD$ bisect AB to obtain E. (See Fig. VIII(a). By the theorem of Pythagoras $EC = \frac{1}{2}\sqrt{5}$. Transfer EC to EF, then

$$AF = \tfrac{1}{2}(1+\sqrt{5}) = \tau \quad \text{and} \quad BF = -\tfrac{1}{2}(1-\sqrt{5}) = -\sigma.$$

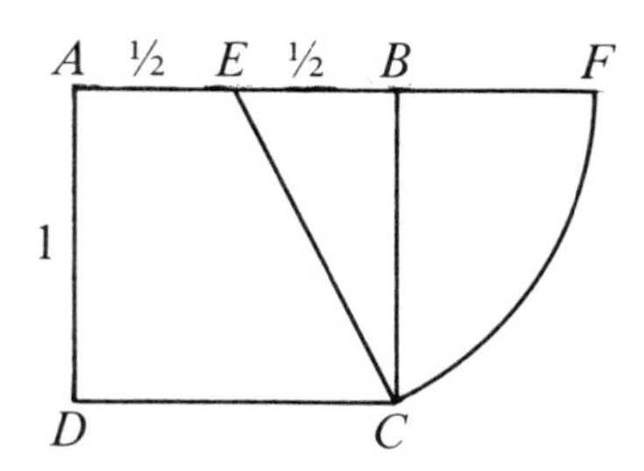

Fig. VIII(a).

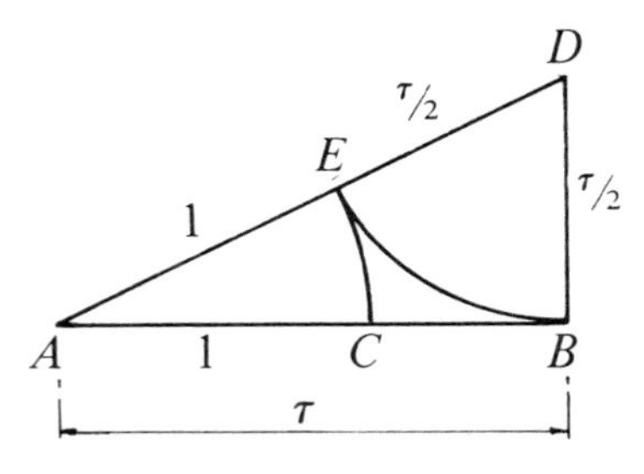

Fig. VIII(b).

It is equally simple to divide a given length $AB = \tau$ in the ratio $1:(\tau - 1)$ (see Fig. VIII(b). If $AB = \tau$, let $BD = \frac{1}{2}\tau$, and let the angle ABD equal 90°. Let $DE = DB$, and $AC = AE$. Then

$$AD = \tfrac{1}{2}\tau\sqrt{5}, \quad \text{and} \quad AC = \tfrac{1}{2}\tau(\sqrt{5} - 1) = 1 \ .$$

3. Let a Golden Rectangle be given (see Fig. IX). Denote the top-left vertex by A.

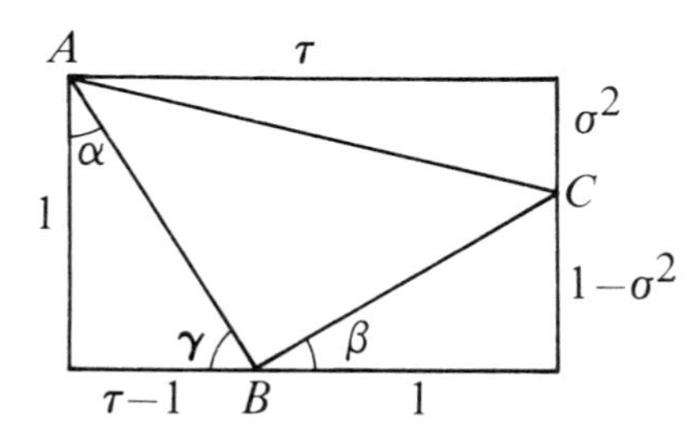

Fig. IX.

Divide the base into lengths $\tau - 1$ and 1, to obtain B; divide the right-hand side into $(\tau - 1)^2 = \sigma^2$ and $1 - (\tau - 1)^2 = 1 - \sigma^2$, to obtain C. Then

$$AB^2 = 1 + \sigma^2 \quad \text{and} \quad BC^2 = 1 + (1 - \sigma^2)^2 \ .$$

Since

$$\sigma^2 = (1 - \sigma^2)^2 \ ,$$

this means $AB = BC$. Observe, also, that

$$(\tau - 1)/1 = \sigma^2/(1 - \sigma^2) \ .$$

(See Hunter, 1963, 1964.)

Denote the angles α and β as in Fig. IX. We have $\tan\alpha = -\sigma$, and $\tan\beta = 1 - \sigma^2 = -\sigma$, hence $\alpha = \beta$ and $\gamma = 90° - \alpha$. Then $\measuredangle ABC = 180° - \beta - \gamma = 180° - \alpha - (90° - \alpha) = 90°$; the triangle ABC is isosceles and right-angled. Its area is $\frac{1}{2}(AB \times BC) = \frac{1}{2}(1 + \sigma^2)$.

The three triangles at the vertices have areas

$$-\tfrac{1}{2}\sigma, \quad \tfrac{1}{2}(1 - \sigma^2), \quad \tfrac{1}{2}\sigma^2\tau = -\sigma/2$$

These are equal.

4. Let an equilateral triangle with sides $1 + \tau = \tau^2$ be given. (See Fig. X.) Divide the

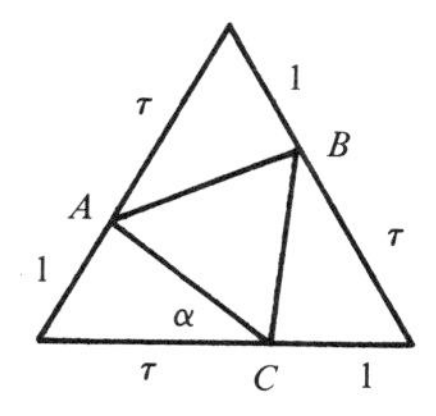

Fig. X.

sides in proportion $1 : \tau$, as in the figure.

The square of the length of a side of the inscribed equilateral triangle is, by the cosine rule of trigonometry,

$$AB^2 = \tau^2 + 1 - 2\cos 60°\ \tau = \tau^2 + 1 - \tau = 2 \ .$$

The area of the inscribed triangle is

$$2 \times \sqrt{3}/4 = \tfrac{1}{2}\sqrt{3}$$

and that of the larger, initial triangle is

$$(AB)^2\sqrt{3}/4 = \tau^4\sqrt{3}/4 \ .$$

5. Consider a semicircle of radius 1 within a square $ABCD$, as in Fig. XI. Connect

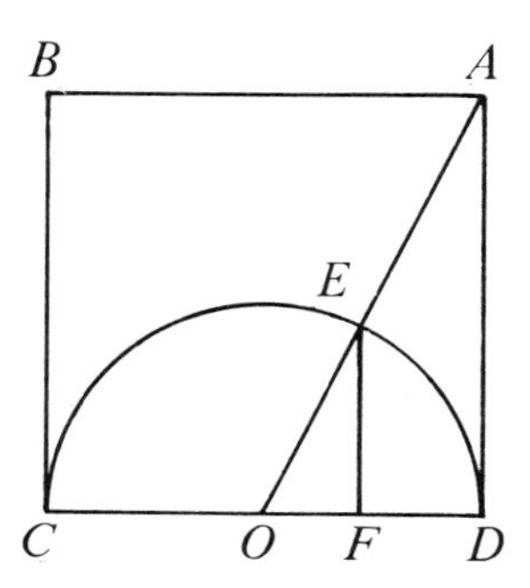

Fig. XI.

the vertex A and the centre of the semicircle by a straight line which intersects the semicircle at E. Draw EF, the perpendicular from E onto the diameter of the semicircle. Then

$$OF/FD = \tfrac{1}{2}\tau \ .$$

Proof. $OE = 1$, $OA = \sqrt{5}$,
$OF/FD = OE/EA = OE/(OA - OE)$

$$= \frac{1}{\sqrt{5}-1} = \frac{\sqrt{5}+1}{4} = \tfrac{1}{2}\tau \ .$$

6. Consider a rectangle whose shorter sides have length 1, and whose longer sides have length $\alpha > 1$. Divide the rectangle into a square of sides 1, and a rectangle of sides 1 and $\alpha - 1$. Treat the latter, residual rectangle in the same way, dividing it into a square of sides $\alpha - 1$, and a rectangle of sides $2 - \alpha$ and $\alpha - 1$. This is, of course, only possible if $\alpha - 1 < 1$, i.e. if $\alpha < 2$.

When $\alpha < 2$, then we can apply a further, analogous step: divide into a square of sides $2-\alpha$ and a rectangle of sides $2\alpha-3$ and $2-\alpha$. For this to be possible, we must have $2-\alpha<\alpha-1$, i.e. $\alpha>3/2$.

Looking at the above inequalities for α, we observe that α is, alternately, larger and smaller than the first convergents of the continuous fraction for τ.

If we carry on in the given fashion, then this continues to be the case. It would appear that for a continuation ad infinitum to be possible, we must have

$$\alpha>F_{n+2}/F_{n+1}\ , \qquad \alpha<F_{n+3}/F_{n+2}\ , \qquad \alpha>F_{n+4}/F_{n+3}$$

and so on. This suggests that we should choose $\alpha=\tau$, and this can be confirmed by observing that when a rectangle $m\times n$ $(m>n)$ is divided into areas $n\times n$ and $(m-n)\times n$, then to ensure continuation without limit, the latter rectangle must be similar to the original one, that is $n/m=(m-n)/n$, or

$$\left(\frac{m}{n}\right)^2=\left(\frac{m}{n}\right)+1, \quad \text{and} \quad \left(\frac{m}{n}\right)>0\ .$$

This means $m/n=\tau$: the original rectangle, and all subsequent ones, must be Golden rectangles.

In Fig. XII we depict the situation, starting with the Golden Rectangle $ABCD$.

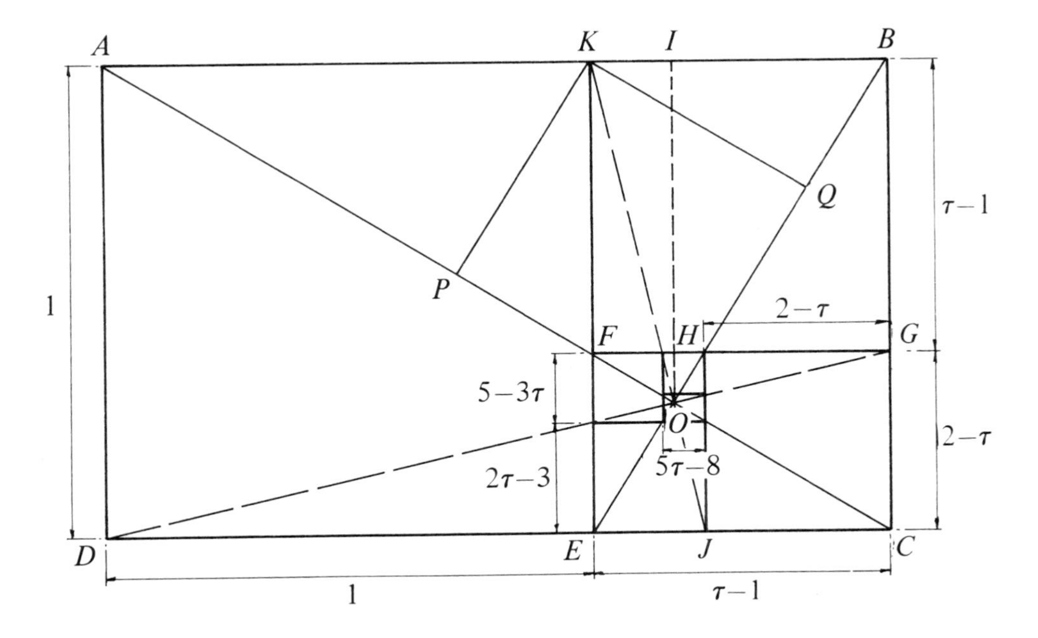

Fig. XII.

We have also drawn the diagonal AC of this rectangle, and the diagonal BE of the rectangle $BCEK$. We denote their intersection by O.

Imagine now that we turn the rectangle $ABCD$ clockwise through a right angle with O as the pivot, and then reduce it in proportion $1/\tau$, with O as fixed point. It will then cover the rectangle $KBCE$. Then we turn this rectangle clockwise through a right angle and contract it again in proportion $1/\tau$, with O as the pivot and fixed point, to cover $CEFG$, and so on. At each of these turns, one diagonal will turn into the other. This shows that the two diagonals intersect at O at a right angle. Every one of the emerging squares will have one of its vertices on one, and the opposite vertex on the other of these diagonals.

Clearly, every square in this process is a gnomon; that is it completes a figure (this time a rectangle) into a similar one.

The successive squares converge to a point, viz. O.

We draw also a perpendicular OI from O onto the side AKB. In Fig. XII we have now three similar triangles, viz.

$$ABC, \ AIO, \ \text{and} \ \ OIB \ .$$

Since $AB/BC = \tau$, we have also $AI/IO = OI/IB = \tau$.

Let AI $= \mu$, then $IO = \mu/\tau$ and hence $IB = \mu/\tau^2$. Also, $AI + IB = AB = \tau$, hence from $\mu + \mu/\tau^2 = \tau$ we conclude that

$$\mu = \frac{\tau^3}{\tau^2 + 1} \ .$$

Thus $AI = \tau^3/(\tau^2 + 1)$ and $IO\tau^2/(\tau^2 + 1)$. Consequently

$$AO = \tau^2/\sqrt{\tau^2 + 1} \quad \text{and} \quad OC = 1/\sqrt{1 + \tau^2} \ .$$

Moreover, again by similarity, $AO/OB = 1/(\tau - 1)$. It follows, that a line OK, which passes through J, bisects the angle AOB, and that a line OG, which passes through D, bisects the angle BOC.

To prove this, imagine a perpendicular KP to be drawn onto AF, and a perpendicular KQ to be drawn onto BH. Then $KPOQ$ is a rectangle, and it has to be proved that it is, in fact, a square, in which diagonals bisect the angles.

Now $KQ/AO = (\tau - 1)/\tau$, and $KP/BO = 1/\tau$. But $AO/BO = 1/(\tau - 1)$, and $KP = KQ$ follows. The second part of our statement is proved in an analogous way.

We imagine now that we introduce a system of polar coordinates, with pole O and base line OD. Consider opposite vertices of the squares, such as $D, K, G, J, \ldots$ Let us take OD for our unit length. Then the polar coordinates of D, K, G, J, and so on are

$$D = (1{,}0), \quad K = (1/\tau, \ \pi/2), \quad G = (1\tau^2, \ \pi), \ J = (1/\tau^3, \ 3\pi/2), \ \ldots$$

The points lie on a curve defined in polar coordinates (r,θ) as

$$r = \tau^{-2\theta/\pi}$$

This is a logarithmic (equiangular) spiral.

7. Let a regular pentagon be given. Its diagonals form a figure called a pentagram; they include a smaller regular pentagon.

Let the sides of the larger pentagon have length 1.

The angles of a regular pentagon are 108°. Therefore (for notation see Fig. XIII)

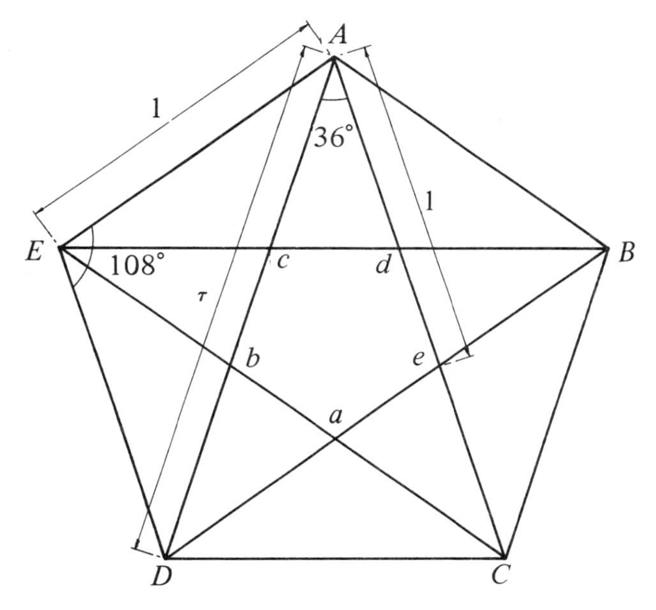

Fig. XIII.

$\measuredangle AdB = 108°$, and since $Ad = dB$, $\measuredangle dAB = 36°$. Using the same argument, we have $\measuredangle cAE = 36°$, and hence $\measuredangle cAd = 108° - 2 \times 36° = 36°$ as well.

$$ADC \text{ is isosceles, hence } \measuredangle ADC = \measuredangle ACD = 72°.$$

Now consider the triangle CbD. By analogy with the computations of angles above, we have $\measuredangle bCD = \measuredangle CAB = 36°$, $\measuredangle bDC = \measuredangle DAB = 72°$, hence CbD is also an isosceles triangle, similar to ADC. (AbC is also a gnomon, in ADC: when omitted, it leaves a figure similar to the original one.)

Now $AC : DC = DC : eC$, but $De = Ae = 1$, therefore

$$AC : 1 = 1 : (AC - 1), \quad \text{and} \quad AC = \tau,\ eC = \tau - 1 = \tau^{-1}\ .$$

Consequently $de = \tau - 2(\tau - 1) = 2 - \tau = \tau^{-2}$.

We have proved: in a regular pentagon with side length 1, the length of diagonal is τ, and the length of a side of the inner pentagon is τ^{-2}. Thus a diagonal, for instance EB, is divided into $Ec = \tau^{-1}$, $cd = \tau^{-2}$, and $dB = \tau^{-1}$. It follows that cos ABE = cos $36° = \frac{1}{2}\tau$. (Hence sin $54° = \frac{1}{2}\tau$, and $\sigma - 2\sin 18°$).

On the other hand, given $AC = 1$, then $Ae = \tau^{-1}$, and the triangle DeC is such that $De = DC = Ae$. Thus a regular pentagon can be constructed with straight edge and compass.

(N.B. C. F. Gauss has proved that a regular pentagon of p sides, p being a prime, can only be constructed with straight edge and compass when p is of the form $2^{2^n} + 1$, but the construction for $n = 1$, i.e. $p = 5$ was already known before his time).

It is also easy to see that the area of a regular pentagon with side length 1 is $5\tau/4\sqrt{3 - \tau}$.

To compute the radius R of the circumcircle of such a pentagon we use again the cosine rule of trigonometry

$$1^2 = R^2 + R^2 - 2R\cos 72° \ ,$$

that is $R^2 = 1/[2(1 - \cos 72°)]$. Now $\cos 72° = \sin 18° = \frac{1}{2}(\tau = 1)$, and therefore $R^2 = 1/(3 - \tau)$.

The inradius r of the pentagon, that is the perpendicular distance of one of its sides from its centre is found from

$$r^2 = R^2 - \tfrac{1}{4} = \tau^2/4(3 - \tau)$$

(where we have used $1 + \tau = \tau^2$).

8. Figure XIV exhibits a portion of Fig. XIII, namely the triangle ACD, with some additional patterns.

The points $ACDba$ repeat those in Fig. XIII. Moreover, we have drawn the median from D in the triangle ADC, the median from b in triangle CbD, and the median from a in the triangle Dab. All these meet in O.

Now imagine the triangle ACD turned clockwise through 108°, with O as the pivot, and then contracting it, with O fixed, in the ratio $1:\tau$. This makes it coincide with CDb. The procedure can be repeated, this time making CDb, after contraction, to coincide with Dba, then the latter (after contraction) with bap, and so on, The successively smaller, but always similar triangles converge to O, while the medians, always passing through O, turn again into medians.

If we denote the distance AO by s (its precise value is of no interest in this context), then the points $A\ C\ D\ b\ a\ p \ldots$ lie on a logarithmic (equiangular) spiral whose equation in polar coordinates reads

$$r = s\tau^{-\theta/(3\pi/5)}$$

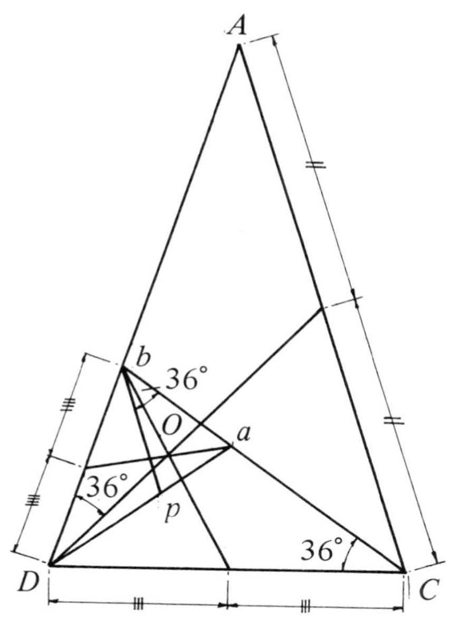

Fig. XIV.

9. Consider a regular decagon with side length 1, denote its centre by O and the vertices by $A, B, \ldots, J$ (see Fig. XV).

If the radius of the circumcircle is denoted by R, then

$$\sin 18° = 1/2R \ .$$

Now $\sin^2 18° = \frac{1}{2}(1 - \cos 36°) = (2 - \tau)/4$, so that $R = 1/\sqrt{2-\tau} = \tau$.

The regular decagon has 35 diagonals,

10 of type I such as AC
10 of type II such as AD
10 of type III such as AE, and
5 of type IV such as AF.

The last mentioned are diameters of the circumcircle.

The diagonals of type I are sides of a pentagon with circumradius τ. We have seen that when the side of a regular pentagon is 1, then its circumcircle has a radius of length $1/\sqrt{3-\tau}$. Hence when the circumradius has length τ, then the side of the corresponding pentagon has length $\tau\sqrt{3-\tau} = \sqrt{2+\tau}$ (about 1.90). This is the length of a diagonal of type I.

The triangle AOD within the pentagon is similar to BAE in Fig. XIII. There the

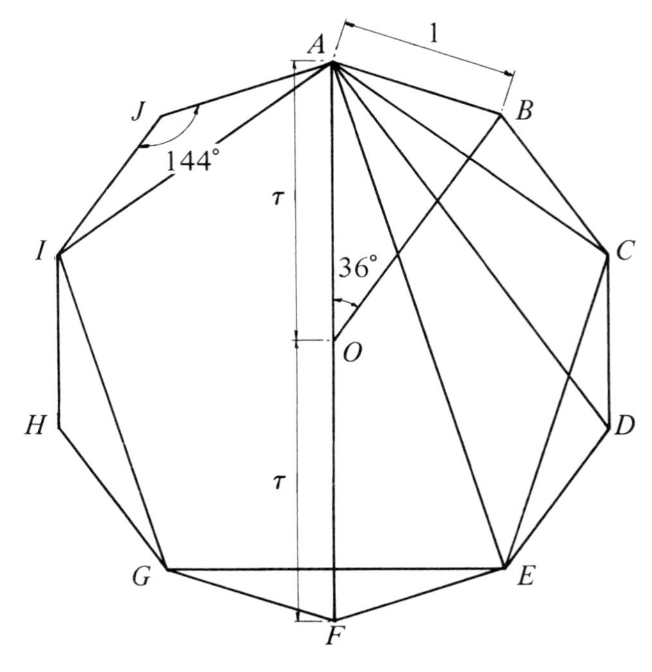

Fig. XV.

length of *Ab* was 1, now the side *AO* has length τ, therefore the length of *AD*, a diagonal of type II, is τ^2 (about 2.62).

AE is a diagonal of the pentagon *ACEGI* with circumradius τ. When this radius was of length $1/\sqrt{3-\tau}$, then the length of the diagonal was τ. Now the radius is τ, therefore the length of AE, a diagonal of type III, is $\tau^2\sqrt{3-\tau} = \tau\sqrt{2+\tau}$ (about 3.08).

Finally, the length of a diameter of the circumcircle is clearly $2R = 2\tau$ (about 3.24).

The inradius of the decagon has length

$$\sqrt{R^2 - \tfrac{1}{4}} = \sqrt{\tau^2 - \tfrac{1}{4}} = \sqrt{\tau + \tfrac{3}{4}} \quad \text{about } 1.43).$$

XIV

The Golden Section in three-dimensional space

Of the five regular platonic solids we shall deal with the octahedron, the dodecahedron, and the icosahedron. They have the following numbers of elements:

	Faces	Edges	Vertices
Octahedron	8	12	6
Dodecahedron	12	30	20
Icosahedron	20	30	12

Consider, first, an octahedron with edges of length $\tau^2\sqrt{2}$. (This is convenient for what follows.) Let the coordinates of the five vertices shown in Fig. XVI be as follows

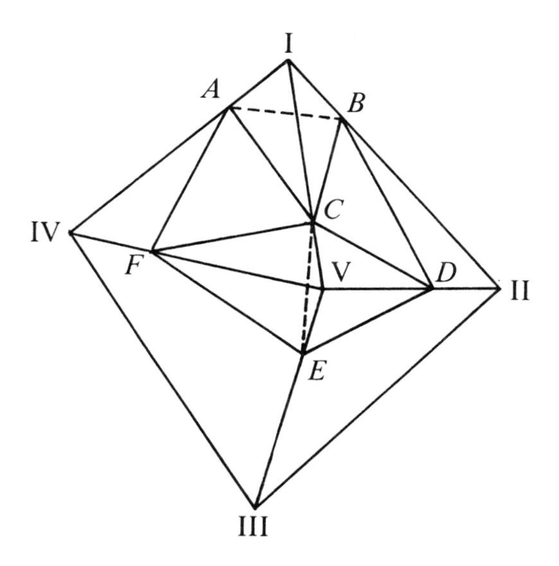

Fig. XVI.

I	II	III	IV	V
$(-\tau^2,0,0)$	$(0,0,\tau^2)$	$(\tau^2,0,0)$	$(0,0,-\tau^2)$	$(0,\tau^2,0)$.

(The sixth vertex $(0,-\tau^2,0)$ lies behind the square (I II III IV) in the figure.)

We divide six of the edges in proportion $1:\tau$, to obtain the points A,B,C,D,E,F, (see figure).

Both triangles ACF and BCD lie on one of the faces of the octahedron, while the lines AB and CE cut into the solid.

We want to prove that the five triangles

$$ABC,\ BDC,\ DEC,\ EFC, \text{ and } FAC$$

are congruent equilateral triangles.

For those triangles which lie on a face of the solid, the computation was done in Chapter XIII, and illustrated in Fig. X. There the side of the larger triangle had length τ^2, and the length of a side of the smaller, inscribed triangle was computed to be $\sqrt{2}$. In the present case the side of the larger triangle — a face of the octahedron — has length $\tau^2\sqrt{2}$, therefore the length of the smaller triangle (i.e. ACF and $DBC, \ldots$) has length 2.

We turn now to one of the other triangles mentioned, for instance ABC. Consider the side AB. It lies on the square (I,II,III,IV). By their construction, A as well as B divide their respective sides of length $\tau^2\sqrt{2}$ in proportion $1:\tau$, hence $AI = BI = \tau^2\sqrt{2}/(1+\tau) = \sqrt{2}$. It follows that $AB = \sqrt{AI^2 + BI^2} = 2 = AC$. Thus we see that the equilateral triangle ABC has also side length 2. That the equilateral triangles EFC and DEC have the same side lengths follows in the same way.

We apply the division in ratio $1:\tau$ to all edges of the octahedron and obtain thus 12 vertices of a polyhedron with 20 faces, all of them equilateral triangles; in fact, we obtain an icosahedron. We have proved: an icosahedron can be inscribed in an octahedron. (Schönemann, 1873).

From Fig. XVI we can compute the coordinates of the vertices of the inscribed icosahedron, with edge length 2. These coordinates are as follows:

A	$-\tau$	0	-1	G	τ	0	-1
B	$-\tau$	0	1	H	τ	0	-1
C	-1	τ	0	I	1	$-\tau$	0
D	0	1	τ	J	0	-1	$-\tau$
E	1	τ	0	K	-1	$-\tau$	0
F	0	1	$-\tau$	L	0	-1	τ

On close examination we find that the set of the 12 vertices can be split up into three sets of four vertices each, forming Golden Rectangles: $AHGB$, $DFJL$, and $CKIE$. The lengths of their sides are 2 and 2τ respectively, and their diagonals have length $2\sqrt{2+\tau} = 2\tau\sqrt{3-\tau}$. Moreover, these three rectangles within the icosahedron are mutually perpendicular; all three pass through the origin of the system of coordinates, and symmetry considerations show that the first rectangle mentioned is perpendicular to the line connecting $(0,0,0) = O$ with $(0,\tau^2,0)$, the second is perpendicular to the line connecting O and $(\tau^2,0,0)$, and the third to the line connecting O and $(0,0,\tau^2)$.

Note: because the centres of the faces of a dodecahedron are vertices of an icosahedron, we can also place four mutually perpendicular Golden Rectangles within a dodecahedron, constructing them out of the set of the centres of the faces of the latter.

We exhibit the vertices of the icosahedron in a diagram (map) in Fig. XVII.

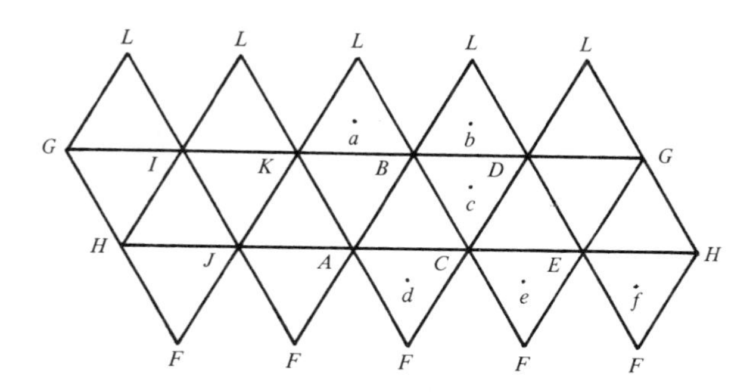

Fig. XVII.

(We shall refer to the points $a, \ldots, f$, the midpoints of their respective triangles, when we deal with the dodecahedron, later in this chapter.)

We are also interested in the lengths of various lines on and within the icosahedron of edge length 2.

There are two types of diagonals, such as LJ, and such as LF. Now LJ is a diagonal of a pentagon with side IJ, hence the length of LJ is 2τ, and LF is a diagonal of one of the inscribed Golden Rectangles, hence the length of **LF** is $2\sqrt{2+\tau} = 2\tau\sqrt{3-\tau}$.

We compute also, for the same icosahedron, the inradius R_i, that is the radius of the sphere which touches all faces, the mid-radius R_m, that is the radius of the sphere which touches all edges, and the circumradius R_c, that is the radius of the sphere which passes through all vertices.

To compute R_i, find the centre of one of the faces of the icosahedron, for instance the centre of the triangle. LKB. Its coordinates are

$$\left(\frac{-1-\tau}{3}, \frac{-1-\tau}{3}, \frac{1+\tau}{3}\right).$$

The distance from the centre O to this point is $\sqrt{(1+\tau)^2/3} = \tau^2/\sqrt{3}$.

R_m connects the centre O and the midpoint of an edge, for instance of AB, that is $(-\tau,0,0)$. Hence $R_m = \tau$.

Finally, R_c is one half of the distance LF, hence $\sqrt{2+\tau} = \tau\sqrt{3-\tau}$.

For convenience, we list here our results for an icosahedron of edge length 1

$$R_c = \tfrac{1}{2}\tau\sqrt{3-\tau}$$

$$R_m = \tfrac{1}{2}\tau$$

$$R_i = \tfrac{1}{2}\tau^2/\sqrt{3}\,.$$

We shall now deal with the dodecahedron in a similar way.

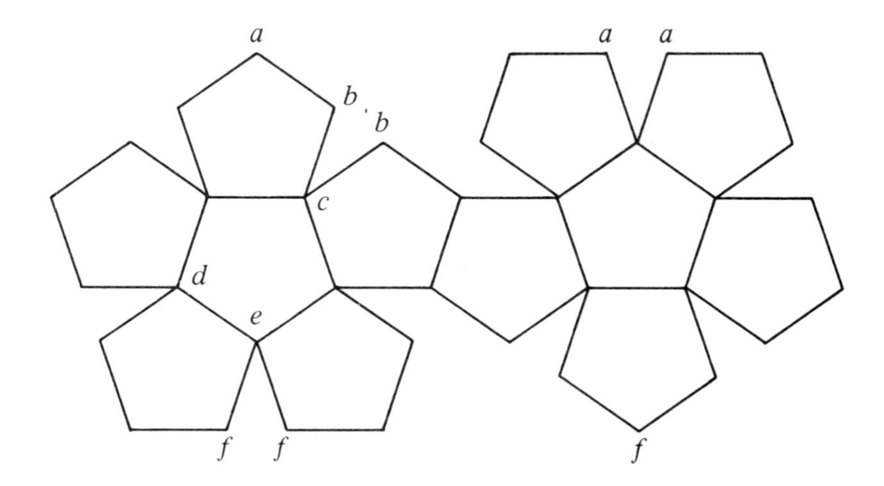

Fig. XVIII.

Let its edges (for instance *ab* in figure XVIII) have length 1. A diagonal of its faces, such as *ac*, has then length τ (see Chapter XIII).

There are three types of diagonals within this solid, such as *ad*, *ae*, and *af*. To compute their lengths, we consider the vertices of the dodecahedron to be the midpoints on the faces of icosahedron, for instance:

a to be the midpoint of	*LKB*, with coordinates	$(-\tau^3/3, -\tau^2/3, \tau^2/3)$
b	*LBD*	$(-\tau/3, 0, \tau^3/3)$
c	*BDC*	$(-\tau^2/3, \tau^2/3, \tau^2/3)$
d	*ACF*	$(-\tau^2/3, \tau^2/3, -\tau^2/3)$
e	*CEF*	$(0, \tau^3/3, -\tau/3)$
f	*EHF*	$(\tau^2/3, \tau^2/3, -\tau^2/3)$.

We find the following distances:

$$ab = 2\tau/3,\ ac = 2\tau^2/3,\ ad = 2\sqrt{2}\tau^2/3,\ ae = \sqrt{2}\tau^3/3,\ af = 2\tau^2/\sqrt{3}\,.$$

However, as we see from *ab*, this refers to a dodecahedron with edge length $2\tau/3$. When the edge length is 1, then we have

$$ab = 1,\ ac = \tau,\ ad = \tau\sqrt{2},\ ae = \tau^2/\sqrt{2},\ af = \tau\sqrt{3}.$$

Now we turn to the radii of the dodecahedron.

The circumradius equals $\tau\sqrt{3}/2$, one half of the length of *af*.

The midradius can be found from the triangle *Oab* (where *O* is the centre of the solid). If *M* is the midpoint of the edge *ab*, then

$$OM = \sqrt{\left(\frac{\tau\sqrt{3}}{2}\right)^2 - \left(\frac{1}{2}\right)^2} = \tau^2/2.$$

To find the length of the inradius, consider the triangle *ONM*, where *N* is the centre of a pentagon face. We want to find the length of *NO*. We have

$$NM^2 = \frac{1}{3-\tau} - \frac{1}{4} = \tau^2/[4(3-\tau)].$$

Hence $NO^2 = OM^2 - NM^2 = \tau^4/4 - \tau^2/[4(3-\tau)] = \tau^4/[4(3-\tau)]$, and *NO*, the inradius whose length we want to find, has length

$$\frac{\tau^2}{2}\frac{1}{\sqrt{3-\tau}}.$$

We have thus found the lengths of the various radii in a dodecahedron of edge length 1, as follows:

Circumradius	$\tau\sqrt{3}/2$
Midradius	$\tau^2/2$
Inradius	$\tau^2/[2\sqrt{3-\tau}]$.

Observe that the ratio of the circumradius to the inradius, $\sqrt{3(3-\tau)}/\tau$, is the same for the icosahedron as it is for the dodecahedron. Consequently, if a dodecahedron and an icosahedron have the same circumsphere, then they have also the same insphere. (In the so-called 14th book of Euclid this remark is attributed to Apollonius of Perga.)

To conclude this chapter, we compute the surface areas and the volumes of these two polytopes.

The surface of an icosahedron with edge length 1 is 20 times the area of one of its faces, hence $5\sqrt{3}$. Its volume is one-third of its surface area, multiplied by the inradius, hence

$$\frac{5\sqrt{3}}{3} \times \frac{\tau^2}{2\sqrt{3}} = \frac{5\tau^2}{6}.$$

The surface of a dodecahedron with edge length 1 is 12 times the area of one of its faces, hence $15\tau/\sqrt{3-\tau}$. Its volume is one-third of its surface, multipled by the inradius, i.e.

$$\frac{5\tau^3}{2(3-\tau)}.$$

We have obtained the following results:

	Solids with edge length 1	
	Icosahedron	Dodecahedron
Surface	$5\sqrt{3}$	$15\tau/\sqrt{3-\tau}$
Volume	$\dfrac{5\tau^5}{6}$	$\dfrac{5\tau^3}{2(3-\tau)}$

These values were computed for solids with edge lengths 1. Now we shall consider an icosahedron–dodecahedron pair with equal inradii. The values above for the two inradii indicate that we can make them equal by multiplying the linear dimensions of the icosahedron by $\sqrt{3}/\sqrt{3-\tau}$. Then its surface will equal $(5 \times 3\sqrt{3})/(3-\tau)$, and its volume will equal

$$\frac{5\tau^2}{6} \times \frac{3\sqrt{3}}{(3-\tau)\sqrt{3-\tau}}.$$

We have obtained

	Enalarged icosahedron	Dodecahedron
Surface	$\dfrac{15\sqrt{3}}{3-\tau}$	$\dfrac{15\tau}{\sqrt{3-\tau}}$
Volume	$\dfrac{5\sqrt{3}\tau^2}{2(3-\tau)\sqrt{(3-\tau)}}$	$\dfrac{5\tau^3}{2(3-\tau)}$.

The ratio of the surfaces of the two solids with equal insphere (and hence equal circumsphere) is

$$\frac{\sqrt{3}}{\tau\sqrt{3-\tau}}$$

and this is also the ratio of the two volumes. As we have pointed out, this is a consequence of the equality of the inradii.

Appendix

1. Congruences

We quote here results from number theory, that are used in the main text, where we did not want to interrupt the flow of an argument by background material.

The numbers we are dealing with here are integers, positive or negative.

If we subtract a multiple of m, say cm from n, then we obtain the 'residue' $r = n - cm$. Whatever the value of c, we write

$$n \equiv r \pmod{m} \ ,$$

pronounced n is congruent to r modulo m.

The smallest non-negative residue is obtained when we divide n by m, assuming $n \geqslant m$. The remainder equals the smallest residue. We shall refer to it as the residue, when no ambiguity is possible.

We have $n \equiv 0 (\text{mod } m)$ if and only if n is divisible, without a remainder, by m.

The largest common factor of s and t is denoted by $(s;t)$, and if $(s;t) = 1$, then we call s and t relatively prime.

If $(s;m) = 1$ and $sa \equiv sa' \pmod{m}$, then $a \equiv a' \pmod{m}$.

The set of numbers which are congruent to t modulo m is called a residue class modulo m. There are m residue classes modulo m, represented by the smallest residues in their respective classes $0,1,\ldots,m-1$.

Euler's Function $\Phi(m)$ denotes the number of the positive integers less than and relatively prime to m. There are $\Phi(m)$ classes of residues prime to m, and any set of $\Phi(m)$ numbers, one from each class, is a 'complete' set.

If $(a_1,\ldots,a_{\Phi(m)})$ is a complete set of residues prime to m, and if $(k;m) = 1$, then

$$(ka_1,\ldots,ka_{\Phi(m)})$$

is also such a complete set.

If $(m,m') = 1$, then $\Phi(mm') = \Phi(m)\,\Phi(m')$. Therefore, to find $\Phi(s)$, we need only know all $\Phi(t)$ where the t are prime power divisors of s.

Of course, when m is prime, then $\Phi(m) = m - 1$.
We have

$$\Phi(p^c) = p^c\,(1 - 1/p)\ .$$

Proof. Of the $p^c - 1$ positive numbers less than p^c, $p^{c-1} - 1$ are multiples of p, and

$$p^c - 1 - (p^{c-1} - 1) = p^c(1 - 1/p)\ .$$

Also, $\Sigma\Phi(d) = m$, the summation extending over all d which divide m.

Fermat's Theorem. Let p be a prime. If $(a;p) = 1$ (which means that p does not divide a) then

$$a^{p-1} \equiv 1 \pmod{p}\ .$$

We prove this as a special case of the

Fermat–Euler Theorem, If $(a;m) = 1$, then $a^{\Phi(m)} \equiv 1$ *(mod m)*.

Proof. Let $x_1, x_2, \ldots, x_{\Phi(m)}$ be a complete set of residues prime to m. If $(a,m) = 1$, then

$$ax_1,\ ax_2, \ldots, ax_{\Phi(m)}$$

is also such a complete set (see above). Hence

$$a^{\Phi(m)}\prod x_i \equiv \prod x_i \pmod{m}$$

where the product extends over the set of residues.

All x_i are relatively prime to m, and so is their product. It follows that

$$a^{\Phi(m)} \equiv\ 1 \pmod{m}\ .$$

When m is a prime, then $a^{p-1} \equiv 1 \pmod{m}$. This is Fermat's (little, not last) Theorem.

2. The Euclidean Algorithm. Continued fractions

Let a and b be two integers, and $b \neq 0$. We want to find $(a;b)$, the largest common divisor of a and b.

Divide a by b, thus

(i) $\qquad a = bq_1 + r_1\ . \qquad 0 \leq r_1 < b$

If $r_1 = 0$, then $(a;b) = b$. Otherwise we continue, dividing b by r_1, thus

(ii) $b = r_1 q_2 + r_2$. $0 \leq r_2 < b$ (If $r_2 = 0$, then $(a;b) = r_1$.)

In this fashion we continue, until we find, eventually

(k) $r_{k-2} = r_{k-1} q_k + r_k$ $(0 \leq r_k < r_{k-1})$

and

(k + 1) $r_{k-1} = r_k q_{k+1} + r_{k+1}$ with $r_{k+1} = 0$,

because the remainders are strictly decreasing.

We have then found $(a;b) = r_k$.

The sequence of equations (i), (ii),. . . can be written as follows:

$$\frac{a}{b} = q_1 + \frac{r_1}{b}$$

$$= q_1 + \frac{1}{\frac{b}{r_1}}$$

$$= q_1 + \frac{1}{q_2 + \frac{r_1}{r_2}}$$

and so on, until

$$\frac{a}{b} = q_1 + \cfrac{1}{q_2 + \cfrac{1}{q_3 + \cfrac{1}{\ddots \; q_k + \cfrac{1}{q_{k+1}}}}} .$$

We have expressed the rational number a/b as a 'continued fraction'.

Let us introduce a more concise notation. For

$$a_0 + \cfrac{1}{a_1 + \cfrac{1}{a_2 + \cfrac{1}{\ddots \, \cfrac{1}{a_N}}}}$$

we write

$$[a_0,a_1,\ldots,a_N]$$

Then

$$[a_0] = a_0 \ , \qquad [a_0,a_1] = \frac{a_1a_0+1}{a_1}$$

and generally

$$[a_0,a_1,\ldots,a_{n+1}] = \left[a_0,a_1,\ldots,a_n + \frac{1}{a_{n+1}} \right] . \tag{*}$$

We call $[a_0,a_1,\ldots,a_n]$, $0 \leqslant n \leqslant N$ the nth convergent of the continued fraction $[a_1,\ldots,a_N]$.

If, moreover, we define

$$p_0 = a_0,\ p_1 = a_1a_0+1,\ldots,p_n = a_np_{n-1}+p_{n-2}\ (2 \leqslant n \leqslant N)$$

and

$$q_0 = 1,\ q_1 = a_1,\ldots,q_n = a_nq_{n-1}+q_{n-2}\ (2 \leqslant n \leqslant N),$$

then we observe that

$$[a_0] = \frac{p_0}{q_0} \ , \qquad [a_0,a_1] = \frac{p_1}{q_1} \ .$$

In general,

$$[a_0,\ a_1\ldots,a_n] = \frac{p_n}{q_n}$$

which we prove by induction, as follows:

Let $[a_0,\ldots,a_n]=p_n/q_n$.

Then, by (*)

$$[a_o,\ldots,a_{n+1}]=\frac{(a_n+1/a_{n+1})p_{n-1}+p_{n-2}}{(a_n+1/a_{n+1})q_{n-1}+q_{n-2}}$$

$$=\frac{a_{n+1}(a_np_{n-1}+p_{n-2})+p_{n-1}}{a_{n+1}(a_nq_{n-1}+q_{n-2})+q_{n-1}}$$

$$=\frac{a_{n+1}p_n+p_{n-1}}{a_{n+1}q_n+q_{n-1}}=\frac{p_{n+1}}{q_{n+1}} \qquad QED$$

From

$$p_nq_{n-1}=(a_np_{n-1}+p_{n-2})q_{n-1}, \text{ and } p_{n-1}q_n=p_{n-1}(a_nq_{n-1}+q_{n-2})$$

we find that

$$p_nq_{n-1}-p_{n-1}q_n=p_{n-2}q_{n-1}-p_{n-1}q_{n-2}=-(p_{n-1}q_{n-2}-p_{n-2}q_{n-1})$$

Starting with $n=1$

$$p_1q_0-p_0q_1=a_1a_0+1-a_0a_1=1$$

we obtain

$$p_2q_1-p_1q_2=-1$$
$$p_3q_2-p_2q_3=1$$
$$\vdots$$
$$p_nq_{n-1}-p_{n-1}q_n=(-1)^{n-1}$$

or

$$\text{(A)} \qquad \frac{p_n}{q_n}-\frac{p_{n-1}}{q_{n-1}}=\frac{(-1)^{n-1}}{q_nq_{n-1}}\ .$$

This means, that for $n\geqslant 2$, p_n/q_n is smaller than p_{n-1}/q_{n-1} when n is even, and larger when n is odd. Also

$$p_nq_{n-2}-p_{n-2}q_n=(a_np_{n-1}+p_{n-2})q_{n-2}-p_{n-2}(a_nq_{n-1}+q_{n-2})$$

$$= a_n(p_{n-1}q_{n-2}) - a_n(p_{n-2}q_{n-1}) = a_n(-1)^n \ ,$$

or

$$\text{(B)} \qquad \frac{p_n}{q_n} - \frac{p_{n-2}}{q_{n-2}} = \frac{a_n(-1)^n}{q_n q_{n-2}} \ .$$

This means that a convergent p_{2n}/q_{2n} increases strictly with increasing n, and that a convergent p_{2n+1}/q_{2n+1} decreases strictly with increasing n.

It can also be shown that

(i) any convergent with an odd subscript is larger than any convergent with an even subscript, and that
(ii) the value of a continued fraction is larger than that of any of its even convergents and smaller than any of its odd convergents, except that it equals its last convergent, whether odd or even.

If $N > 1$, $n > 0$, then

$$\frac{p_N}{q_N} q_n - p_n$$

decreases steadily in absolute value as n increases, and

$$\left| \frac{p_N}{q_N} - \frac{p_n}{q_n} \right| = \frac{1}{q_n q_{n+1}} < \frac{1}{q_n^2}$$

for $n < N - 1$, and

$$\left| \frac{p_N}{q_N} - \frac{p_{N-1}}{q_{N-1}} \right| = \frac{1}{q_N q_{N-1}} < \frac{1}{q_{N-1}^2} \ .$$

Example:

$a = 18$, $b = 5$.

$$\frac{18}{5} = 3 + \cfrac{1}{1 + \cfrac{1}{1 + \frac{1}{2}}} \ .$$

$a_0 = 3$, $a_1 = 1$, $a_2 = 1$, $a_3 = 2$.

$p_0 = 3$, $p_1 = 4$, $p_2 = 7$, $p_3 = 18$

$$q_0 = 1,\ q_1 = 1,\ q_2 = 2,\ q_3 = 5.$$

The convergents are 3, 4, 3.5, 3.6.

$$\begin{array}{lll}
\text{(A)} & 3.6 - 3.5 = \ \ 0.1 & 3.6 > 3.5 \\
 & 3.5 - 4 \ = -0.5 & 3.5 < 4 \\
 & 4 \ - 3 \ = \ \ 1 & 4 \ > 3 \\
\text{(B)} & 3.6 - 4 \ = -0.4 & 4 \ > 3.6 \\
 & 3.5 - 3 \ = \ \ 0.5 & 3 \ < 3.5 \\
 & \multicolumn{2}{l}{4 > 3,\ 4 > 3.5,\ 3.6 > 3,\ 3.6 > 3.5,\ 3.6 < 4.}
\end{array}$$

$$|3.6 - 3| < \frac{1}{1},\ |3.6 - 4| < \frac{1}{2},\ |3.6 - 3.5| = < \frac{1}{4}.$$

We have seen how a ratio of two integers, i.e. a rational number, can be expressed as a finite continued fraction. If all a_i ($i = 1, \ldots, N$) are positive integers (a_0 may be zero or a negative integer), then the continued fraction is called 'simple'.

We shall now consider infinite simple continued fractions. All such continued fractions are convergent; that is, their convergents tend to a limit.

The rules about decrease and increase of convergents, which we have described in the finite case, apply equally in the infinite case. Every irrational number can be expressed in just one way as an infinite continued fraction.

We call an infinite continued fraction 'periodic', if $a_n = a_{n+k}$ for fixed k (>0) and all n larger than some given integer.

Theorem. A periodic continued fraction is an irrational root of a quadratic equation with integer coefficients.

For the proof, see Hardy and Wright (1960), Theorems 176 and 177.

Example.

$$[a, a, \ldots, a]$$

$$x = a + \cfrac{1}{a + \cfrac{1}{a + \cfrac{1}{\ddots}}}$$

Clearly, $x = a + 1/x$, that is $x^2 - ax - 1 = 0$.

The two roots of this equation are $\frac{1}{2}(a + \sqrt{a^2 + 4}) = x_1$ and $\frac{1}{2}(a - \sqrt{a^2 + 4}) = x_2$. Because the value of the continued fraction is positive, it must equal x_1, since $x_2 < 0$.

We have now

$$p_0 = a = x_1 + x_2 = \frac{x_1^2 - x_2^2}{x_1 - x_2}$$

$$p_1 = ap_0 + 1 = \frac{ax_1^2 - ax_2^2 + x_1 - x_2}{x_1 - x_2} = \frac{x_1^3 - x_2^3}{x_1 - x_2}$$

and generally, by induction

$$p_n = \frac{x_1^{n+2} - x_2^{n+2}}{x_1 - x_2} .$$

In precisely the same way we find

$$q_n = \frac{x_1^{n+1} - x_2^{n+1}}{x_1 - x_2} .$$

It follows that the nth convergent p_n/q_n equals

$$\frac{x_1^{n+2} - x_2^{n+2}}{x_1^{n+1} - x_2^{n+1}} .$$

With increasing n the convergents tend to

$$x_1^{n+2}/x_1^{n+1} = x_1$$

because

$$|x_2| = \left|\frac{a - \sqrt{a^2+4}}{2}\right| < 1 .$$

3. Quadratic residues. Quadratic reciprocity

We are concerned with the quadratic congruence $x_2 \equiv a (\bmod p)$, where p is an odd prime and a an integer, not a multiple of p.

If this congruence has a solution x, then we say that a is a quadratic residue modulo p.

The quadratic residues of p are numbers which are congruent, modulo m, to one of the numbers

$$1^2, 2^2, \ldots, (p-1)^2.$$

They are congruent in pairs, because $(p-x)^2 = p^2 - 2px + x^2 \equiv x^2 (\bmod p)$, and there are no other congruences amongst them, because if two out of the numbers

$1,2,\ldots,\frac{1}{2}(p-1)$, say s and t, satisfy $s^2 \equiv t^2 \pmod{p}$, then $(s+t)(s-t) \equiv 0 \pmod{p}$, and $s \equiv -t$ is impossible (we are dealing with positive values) so that $s = t$.

Hence, half the numbers $1\text{-}x \leqslant p \leqslant 1$ are quadratic residues, and the other half must be quadratic non-residues. Each of these two sets consists of $\frac{1}{2}(p-1)$ numbers.

Consider, once more, the set

$$1,\ldots,p-1.$$

and let us group them into $\frac{1}{2}(p-1)$ pairs. Let one of the pairs be $(x_1, p-x_1)$ where

$$x_1^2 \equiv a (\text{mod } p), \text{ so that}$$

$$x_1(p-x_1) \equiv -a (\text{mod } p)$$

and let the other pairs be (x_2, x_2') such that $(x_2 x_2') \equiv a \pmod{p}$. Of course, $x_2 \neq x_1$.

Now look at the product $1.2.3.\ldots.(p-1) = (p-1)!$. The grouping shows that

(i) $$(p-1)! \equiv (-a)a^{\frac{1}{2}(p-3)} \pmod{p} \equiv -a^{\frac{1}{2}(p-1)} \pmod{p}.$$

We obtain a second formula for $(p-1)!$ as follows.

Let a be a quadratic non-residue modulo p. Then

$$x^2 \not\equiv a \pmod{p}.$$

We can again group the numbers of the set $1,2,\ldots,p-1$ into pairs, this time in such a way that each pair contains numbers whose product is not congruent to a, modulo p. Then

(ii) $$(p-1)! \equiv a^{\frac{1}{2}(p-1)} \pmod{p}.$$

It is convenient to introduce the Legendre symbol $\left(\frac{a}{p}\right)$, defined to be equal to 1, if a is a quadratic residue of p, and equal to -1, if a is a quadratic non-residue modulo p.

(i) and (ii) can then both be expressed by writing

(iii) $$(p-1)! \equiv -\left(\frac{a}{b}\right) a^{\frac{1}{2}(p-1)} \pmod{p}.$$

If in (iii) we set $a = 1$, which is a quadratic residue of any p, we obtain

(iv) $(p-1)! \equiv -\left(\frac{1}{p}\right) \pmod{p} \equiv -1 \pmod{p}$.

In the literature this is called Wilson's Theorem.

We compare (iii) and (iv) and find that

$$\left(\frac{a}{b}\right) \equiv a^{\frac{1}{2}(p-1)} \pmod{p}$$

because $1/\left(\frac{a}{p}\right) = \left(\frac{a}{p}\right)$, (whether $\left(\frac{a}{p}\right)$ equals 1 or -1) and therefore

$$\left(\frac{ab}{p}\right) = \left(\frac{a}{p}\right)\left(\frac{b}{p}\right) .$$

In particular,

$$\left(\frac{-1}{p}\right) \equiv (-1)^{\frac{1}{2}(p-1)} \pmod{p} .$$

This means, that

- -1 is a quadratic residue modulo p if $\frac{1}{2}(p-1)$ is even, that is if p is of the form $4k+1$, and
- -1 is a quadratic non-residue modulo p, if $\frac{1}{2}(p-1)$ is odd, that is if p is of the form $4k+3$.

The Law of Quadratic Reciprocity. This law, stated concisely, says that

$$\left(\frac{p}{q}\right)\left(\frac{q}{p}\right) = (-1)^{\frac{1}{2}(p-1)\frac{1}{2}(q-1)} .$$

More explicitly, this means that

$\left(\frac{p}{q}\right) = -\left(\frac{q}{p}\right)$ if p as well as q are of the form $4k+3$, while otherwise

$$\left(\frac{p}{q}\right) = \left(\frac{q}{p}\right) .$$

In other words, p is a quadratic residue of q if and only if q is a quadratic residue of p, provided

$$\frac{p-1}{2}\frac{q-1}{2} \text{ is even.}$$

This theorem was first stated by Legendre, and the first to give a rigorous proof was Gauss. He added later two more proofs. For these, and numerous other proofs we must refer the reader to Hardy and Wright (1980).

The Law of Quadratic Reciprocity can be used to find primes modulo which a given a, not a multiple of p, is a quadratic residue.

Examples.
Let $p = 5t \pm 1$. Then

$$\left(\frac{5}{p}\right) = \left(\frac{p}{5}\right) = \left(\frac{5t+1}{5}\right) = \left(\frac{1}{5}\right) = 1$$

and $$\left(\frac{5t-1}{5}\right) = \left(\frac{4}{5}\right) = 1$$

since $1^2 \equiv 1 (\text{mod } 5)$, $2^2 \equiv 4 (\text{mod } 5)$, $3^2 \equiv 4 (\text{mod } 5)$, $4^2 \equiv 1 (\text{mod } 5)$. Thus 5 is a quadratic residue modulo primes of the form $5t \pm 1$, and a quadratic non-residue modulo primes of the form $5t \pm 2$,

$$\left(\frac{5}{5t \pm 1}\right) = 1, \quad \left(\frac{5}{5t \pm 2}\right) = -1 .$$

We have seen that $\left(\frac{-1}{p}\right) = 1$ if p is of the form $4k+1$, and -1 if p is of the form $4k+3$.

This enables us to find the primes modulo which -5 is a quadratic residue. We have

$$\left(\frac{-5}{p}\right) = \left(\frac{5}{p}\right)\left(\frac{-1}{p}\right)$$

and this equals 1 if either

$$\left(\frac{5}{p}\right) = 1 \text{ and } \left(\frac{-1}{p}\right) = 1$$

or if

$$\left(\frac{5}{p}\right) = -1 \text{ and } \left(\frac{-1}{p}\right) = -1.$$

As we have seen, this means that -5 is a quadratic residue modulo p
if (i) either $p = 10n \pm 1$ and also $4k + 1$, or
if (ii) either $p = 10n \pm 3$ and also $4k + 3$.
Therefore p must have one of the forms

$$20m + 1,\ 20m + 3,\ 20m + 7, \text{ or } 20m + 9.$$

For instance,

$x^2 \equiv -5 \pmod{21}$ has the solution $x = 4$ (amongst others)
$x^2 \equiv -5 \pmod{3}$ has the solution $x = 2$
$x^2 \equiv -5 \pmod{7}$ has the solution $x = 3$
$x^2 \equiv -5 \pmod{9}$ has the solution $x = 2$.

4. Binomial coefficients

The binomial coefficient $\binom{n}{m}$ is defined, for non-negative integers n and m, $n \geqslant m$, as

$$\binom{n}{m} = \frac{n!}{m!(n-m)!} \tag{A4.1}$$

where $n!$ is the factorial $1.2.\ldots.n$, and $0!$ is defined as 1. Hence $\binom{n}{0} = 1$ for all n. Also,

$$\binom{n}{m} = \binom{n}{n-m}$$

for all n and m. In particular

$$\binom{n}{1} = \binom{n}{n-1} = n.$$

In spite of the appearance of (A4.1), binomial coefficients are integers. To see this, observe first that

$$\binom{n}{m-1} + \binom{n}{m} \binom{n+1}{m} \text{ for } m \geqslant 1. \qquad \text{(A.4.2)}$$

Proof.

$$\begin{aligned} &\frac{n!}{(m-1)!(n-m+1)!} + \frac{n!}{m!(n-m)!} = \frac{n!}{(n-1)!(n-m)!}\left[\frac{1}{n-m+1} + \frac{1}{m}\right] \\ &= \frac{n!}{(m-1)!(n-m)!}\,\frac{n+1}{m(n-m+1)} = \frac{(n+1!}{m!(n-m+1)!} \text{ QED} \end{aligned}$$

$\binom{n}{0}$ and $\binom{n}{1}$ are obviously integers, and we can exhibit the binomial coefficients in the shape of the 'Pascal Triangle'

$$\begin{array}{ccccc} & \binom{1}{0} & & \binom{1}{1} & \\ \binom{2}{0} & & \binom{2}{1} & & \binom{2}{2} \end{array}$$

and so on, where every entry is the sum of the two nearest entries above it, thus:

$$\begin{array}{ccccccccccc} & & & & 1 & & 1 & & & & \\ & & & 1 & & 2 & & 1 & & & \\ & & 1 & & 3 & & 3 & & 1 & & \\ & 1 & & 4 & & 6 & & 4 & & 1 & \\ 1 & & 5 & & 10 & & 10 & & 5 & & 1 \\ \vdots & & & & & & & & & & \end{array}$$

Clearly, all entries are integers.

The Binomial Theorem. Let n be a positive integer. Then

$$(a+b)^n = a^n + \binom{n}{1}a^{n-1}b + \binom{n}{2}a^{n-2}b^2 + \ldots + \binom{n}{n-1}ab^{n-1} + b^n. \tag{A.4.3}$$

We prove this theorem by induction. It holds, trivially, for $n=1$. Assume that it holds for $1,2,\ldots,n$. Then

$$(a+b)^{n+1} = (a+b)^n(a+b) = a\ a^n + a\binom{n}{1}a^{n-1}b + \ldots + a\binom{n}{n}b^n$$

$$+ b\ a^n + b\binom{n}{1}a^{n-1}b + \ldots + b\binom{n}{n}b^n$$

$$= a^{n+1} + \binom{n+1}{1}a^n b + \binom{n+1}{2}a^{n-1}b^2 + \ldots + \binom{n+1}{n}ab^n + b^{n+1}$$

where we have made use of (A.4.2).

We prove now

$$\binom{n+m}{k} = \sum_{i=0}^{k}\binom{n}{i}\binom{m}{k-i}\ . \tag{A.4.3}$$

$(a+b)^{n+m} = (a+b)^n(a+b)^m$, that is from (A.4.3)

$$\sum_{k=0}^{n+m}\binom{n+m}{k}a_k b^{n+m+k} = \sum_{i=0}^{n}\sum_{j=0}^{m}\binom{n}{i}\binom{m}{j}a^{i+j}b^{n+m-i-j}.$$

Now

$$a^{i+j}b^{n+m-i-j} = a^k b^{n+m-k} \text{ if } j=k-1.$$

The coefficient of $a^k b^{n+m-k}$ on one side is $\binom{n+m}{k}$, and on the other side it is $\sum_{i=0}^{k}\binom{n}{i}\binom{m}{k-i}$.

Factors of binomial coefficients. Let p be a prime.

$\binom{p}{n}=\frac{p!}{n!(p-n)!}$ is an integer, hence $n!$ is a factor of $\frac{p!}{(p-n)!}$.

If $1\leqslant n\leqslant p-1$, then $n!$ is relatively prime to p, and must therefore be a factor of $\frac{(p-1)!}{(p-n)!}$.

It follows that $p\frac{(p-1)!}{n!(p-n)!}$ is divisible by p, that is

$$\binom{p}{n}\equiv 0 \pmod p \text{ for } 1\leqslant n\leqslant p-1. \tag{A.4.5}$$

We have $\binom{p-1}{1}\equiv(-1) \pmod p$ and, using (A.4.2) and (A.4.5)

$$\binom{p-1}{2}=\binom{p}{2}-\binom{p-1}{1}\equiv -(-1)=(-1)^2 \pmod p.$$

By induction,

$$\binom{p-1}{n}\equiv(-1)^n \pmod p \ (1\leqslant n\leqslant p-1) \tag{A.4. 6}$$

Moreover,

$$\binom{p}{n-1}\equiv 0 \pmod p \text{ and } \binom{p}{n}\equiv 0 \pmod p \ (2\leqslant n\leqslant p-1)$$

and thus, because of (A.4.2)

$$\binom{p+1}{n}\equiv 0 \pmod p \ (2\leqslant n\leqslant p-1) \tag{A.4.7}$$

5. Difference equations

Consider the recurrence (or difference equation)

$$u_{n+k}=a_1u_{n+k-1}+a_2u_{n+k-2}+\ldots+a_ku_n. \tag{A.5.1}$$

We assume that u_t has the form x^t, then

$$x^{n+k} - a_1x^{n+k-1} - a_2x^{n+k-2} - \ldots - a_ku^n = 0. \quad \text{(A.5.2)}$$

This polynomial equation, the 'characteristic equation' of (A.4.1), has the root $x = 0$, which leads to the sequence 0,0,. . ., of no interest to us. Therefore we assume $x \neq 0$ and divide (A.4.2) by x^n, to obtain

$$f(x) = x_k - a_1x^{k-1} - a_rx^{k-2} - \ldots - a_k = 0.$$

This equation has k roots, say $x_1, \ldots, x_k$. Assume, to begin with, that they are all distinct.

Let $u_n = x_i^n$ $(i = 1,2,\ldots k)$ be solutions of (A.4.1); then

$$u_n = \alpha_1x_1^n + \alpha_2x_2^n + \ldots + \alpha_kx_k^n$$

is also a solution. The coefficients α_i depend on the first k values of the recurrence

$$u_0, u_1, \ldots, u_{k-1}$$

We call the set of these values the 'seed'.

The coefficients α_i are computed from

$$\begin{array}{lllll} u_o & = \alpha_1 & + \alpha_2 & + \ldots & + \alpha_k \\ u_1 & = \alpha_1x_1 & + \alpha_2x_2 & + \ldots & + \alpha_kx_k \\ \vdots & & & & \\ u_{k-1} & = \alpha_1x_1^{k-1} & + \alpha_2x_2^{k-1} & + \ldots & + \alpha_kx_k^{k-1} \end{array}$$

The solution of this system of linear equations depends on the Vandermonde determinant

$$\begin{vmatrix} 1 & 1 & \ldots & 1 \\ x_1 & x_2 & \ldots & x_k \\ \vdots & & & \\ x_1^{k-1} & x_2^{k-1} & \ldots & x_k^{k-1} \end{vmatrix} = \Pi(x_i - x_k) \quad ,$$

where the product extends over all pairs of subscripts (i,j), $j > i$.

The determinant is different from 0 if and only if all x_i are distinct.

When x_0, say, is a t-fold root of the characteristic equation, then

$$u_n = nx_0^n, \; = \; n^2x_0^n, \; \ldots \; = n^{t-1}x_0^n$$

are also solutions of (A.5.1). We prove this for $t=2$. Then we have to show that nx_0^n is also a solution. Substituting into (A.5.1), we have

$$(n+2)x_0^{n+2}=a_1(n+1)x_0^{n+1}+a_2nx_0^n. \tag{A.5.4}$$

Multiply (A.5.1) by n and subtract it from (A.5.4), to obtain

$$2x_0^{n+2}=a_1x_0^{n+1}.$$

Now if x_0 is a double root of $f(x)=0$, then it is also a root of

$$\frac{\mathrm{d}f(x)}{\mathrm{d}x}=0$$

that is of $2x-a_1=0$.

This completes the proof. If we have a t-fold root, $t>2$, then the proof proceeds in an analogous manner.

We have thus again altogether k different solutions, and any linear combination is also a solution. The coefficients of such a linear combination depend once more on the seed.

Example. Let the difference equation be

$$u_{n+3}=4u_{n+2}-5u_{n+1}+2u_n\ .$$

The characteristic equation

$$f(x)=x^3-4x^2+5x-2=0$$

has the roots $x=1$ (twice) and $x=2$.

We must therefore construct the equation

$$u_n=\alpha_1 1^n+\alpha_2 n1^n+a_3 2^n$$

such that the seed is obtained.

When the seed is (0,0,1), this means

$$\alpha_1=-1,\ \alpha_2=-1,\ \alpha_3=1\ .$$

The sequence defined by the recurrence equation is

0 0 1 4 11 26 ...

and indeed

$$u_3 = -1-3+8 = 4$$
$$u_4 = -1-4+16 = 11$$
$$u_5 = -1-5+32 = 26$$

Thus the seed determines the coefficients α_i. Conversely, the coefficients determine the seed.

When $k=3$, then it is easy to find the seed which corresponds to

$$\alpha_1 = \alpha_2 = \alpha_3 = 1 \ .$$

In this case

$$x_1^3 - a_1 x^2 - a_2 x - a_3 = (x-r_1)(x-r_2)(x-r_3) \ ,$$

so that

$$r_1 + r_2 + r_3 = a_1 \ \text{ and } \ r_1r_2 + r_1r_3 + r_2r_3 = -a_2 \ .$$

(Also $r_1r_2r_3 = -a_3$, but this is irrelevant here.)
Therefore

$$r_1^0 + r_2^0 + r_3^0 = 3$$
$$r_1 + r_2 + r_3 = a_1$$
$$r_1^2 + r_2^2 + r_3^2 = (r_1+r_2+r_3)^2 - 2(r_1r_2 + r_1r_3 + r_2r_3) = a_1^2 + 2a_2.$$

For instance, when the characteristic equation is

$$x_3 - x - 1 = 0, \text{ that is } a_1 = 0, \ a_2 = 1, \ a_3 = 1 \ ,$$

then the seed which leads to $\alpha_1 = \alpha_2 = \alpha_3 = 1$ will be $(3, 0, 2)$. (In fact, the roots are, in this example

$$1.325, \ -0.6625 + 0.5525i, \ -0.6625 - 0.5525i.)$$

6. Linear equations and determinants

Consider the system of n linear equations in n unknowns

$$a_{11}x_1 + a_{12}x_2 + \ldots + a_{1n}x_n = b_1$$
$$a_{21}x_1 + a_{22}x_2 + \ldots + a_{2n}x_n = b_2$$
$$\vdots$$

$$a_{n1}x_1 + a_{n2}x_2 + \ldots + a_{nn}x_n = b_n \ .$$

The a_{ij} and the b_i are given, and we want to find values $x_1, x_2, \ldots, x_n$ which solve the system, that is satisfy the equations

This could, in principle, be done by expressing x_1 in terms of the a_{1j} ($j = 1, \ldots, n$) and b_1, substituting the result into the next equation, using this equation to find x_2, and continuing in this way until we find, finally, x_n. Retracing our steps, we then find $x_{n-1}, x_{n-2}, \ldots, x_1$.

Example. $n = 2$.

$$3x_1 + 2x_2 = 7$$
$$x_1 + 4x_2 = 9 \ .$$

We find, first, $x_1 = (7 - 2x_2)/3$, and substitute x_1 into the second equation,

$$(7 - 2x_2)/3 + 4x_2 = 9, \text{ that is } x_2 = 2.$$

Then, $x_1 = (7 - 2x_2)/3 = 1$.

This method can break down, though, as in the next example.

$$3x_1 + 12x_2 = 7$$
$$x_1 + \ \ 4x_2 = 9.$$

We compute

$$x_1 = (7 - 12x_2)/3, \qquad ((7 - 12x_2)/3 + 4x_2 = 9 \ ,$$

that is $7/3 = 9$, which is impossible.

On reflection, we could have seen to begin with that the system is contradictory. The left-hand side of the first equation is three times that of the second equation, but this is not so of the right-hand sides.

However, consider

$$3x_1 + 12x_2 = 27$$
$$x_1 + \ \ 4x_2 = \ \ 9.$$

Now

$$x_1 = (27 - 12x_2)/3 \qquad 9 - 4x_2 + 4x_2 = 9,$$

which is not a contradiction. But what are the values of x_1 and of x_2? All we know about them is that $x_1 = 9 - 4x_2$, so that we have an infinity of pairs which solve the system.

To obtain more general insight into these matters, we proceed as above, but with the general 2 by 2 system

$$a_{11}x_1 + a_{12}x_2 = b_1, \qquad a_{21}x_1 + a_{22}x_2 = b_2.$$

The solution is now

$$x_1 = \frac{b_1a_{22} - b_2a_{12}}{a_{11}a_{22} - a_{12}a_{21}}, \qquad x_2 = \frac{b_2a_{11} - b_1a_{21}}{a_{11}a_{22} - a_{12}a_{21}}.$$

We see now that we obtain a unique solution if and only if

$$a_{11}a_{22} - a_{12}a_{21} \neq 0.$$

We call $a_{11}a_{22} - a_{12}a_{21}$ the determinant of the system of two linear equations in two unknowns and write it

$$\begin{vmatrix} a_{11} & a_{12} \\ a_{21} & a_{22} \end{vmatrix} = D, \text{ say}$$

The a_{ij} appear in the pattern in which they appeared in the system to be solved.

We look now at the system of n linear equations, with which we started. The determinant of such a system is written

$$\begin{vmatrix} a_{11} & a_{12} & \dots & a_{1n} \\ a_{21} & a_{22} & \dots & a_{2n} \\ & & & \\ a_{n1} & a_{n2} & \dots & a_{nn} \end{vmatrix} = D.$$

The value of such a determinant can, in principle, be found recursively, by using the recurrence

$$D = a_{11}\begin{vmatrix} a_{22} & \dots & a_{2n} \\ & & \\ a_{n2} & \dots & a_{nn} \end{vmatrix} - a_{12}\begin{vmatrix} a_{21} & \dots & a_{2n} \\ & & \\ a_{n1} & \dots & a_{nn} \end{vmatrix} \pm \dots (-1)^{n-1} a_{1n}\begin{vmatrix} a_{21} & \dots & a_{2n-1} \\ & & \\ a_{n1} & \dots & a_{nn-1} \end{vmatrix}$$

but in practice advantage is taken of the following rules:

(i) Multiplication of a row, or of a column, by a constant multiplies the value of the determinant by that same constant.

(ii) Interchange of two rows, or of two columns, multiplies the value of the determinant by -1.

(ii) Addition of a multiple of a row (or of a column) to another row (or to another column) does not change the value of the determinat.

Cramer's rule states that the solution of the system (1) reads

$$x_1 = \begin{vmatrix} b_1 & a_{12} & \dots & a_{1n} \\ b_2 & a_{22} & \dots & a_{2n} \\ & & & \\ b_n & a_{n2} & \dots & a_{nn} \end{vmatrix} \Big/ D \quad x_2 = \begin{vmatrix} a_{11} & b_{12} & \dots & a_{1n} \\ a_{21} & b_{22} & \dots & a_{2n} \\ & & & \\ a_{n1} & b_{n2} & \dots & a_{nn} \end{vmatrix} \Big/ D$$

$$x_n = \begin{vmatrix} a_{11} & a_{12} & \dots & b_1 \\ a_{21} & a_{22} & \dots & b_2 \\ & & & \\ a_{n1} & a_{n2} & \dots & b_n \end{vmatrix} \Big/ D$$

The system has a unique solution if and only if $D \neq 0$.

Example.

$$\begin{aligned} 3x_1 + x_2 + 2x_3 &= 13 \\ x_1 + 3x_2 + 5x_3 &= 20 \\ 2x_1 + 3x_2 + x_3 &= 10 \end{aligned}$$

$$D = \begin{vmatrix} 3 & 1 & 2 \\ 1 & 3 & 5 \\ 2 & 3 & 1 \end{vmatrix} = -33 \ .$$

$$x_1 = \begin{vmatrix} 13 & 1 & 2 \\ 20 & 3 & 5 \\ 10 & 3 & 1 \end{vmatrix} \Big/ (-33) = 2$$

$$x_2 = \begin{vmatrix} 3 & 13 & 2 \\ 1 & 20 & 5 \\ 2 & 10 & 1 \end{vmatrix} \Big/ (-33) = 1$$

$$x_3 = \begin{vmatrix} 3 & 1 & 13 \\ 1 & 3 & 20 \\ 2 & 3 & 10 \end{vmatrix} \Big/ (-33) = 3 \ .$$

List of formulae

(1) $F_{n+2} = F_{n+1} + F_n$

(2) $F_{-n} = (-1)^{n+1} F_n$

(3) $G_{n+2} = G_{n+1} + G_n$

(4) $L_{-n} = (-1)^n L_n$

(5) $L_{n-1} + L_{n+1} = 5F_n$

(6) $F_{n-1} + F_{n+1} = L_n$

(7a) $F_{n+2} - F_{n-2} = L_n$

(7b) $F_n + L_n = 2F_{n+1}$

(8) $G_{n+m} = F_{m-1} G_n + F_m G_{n+1}$

(9) $G_{n-m} = (-1)^m (F_{m+1} G_n - F_m G_{n+1})$

(10a) $G_{n+m} + (-1)^m G_{n-m} = L_m G_n$

(10b) $G_{n+m} - (-1)^m G_{n-m} = F_m (G_{n-1} + G_{n+1})$

(11) $F_{n+1}^2 + F_n^2 = F_{2n+1}$

(12) $F_{n+1}^2 - F_n^2 = F_{n+2} F_{n-1}$

(13) $F_{2n} = F_n L_n$

(14) $F_{n+1}L_{n+1} - F_n L_n = F_{2n+1}$

(15a) $F_{n+m} + (-1)^m F_{n-m} = L_m F_n$

(15b) $F_{n+m} - (-1)^m F_{n-m} = F_m L_n$

(16a) $L_m F_n + L_n F_m = 2F_{n+m}$

(16b) $F_n L_m - L_n F_m = (-1)^m 2F_{n-m}$

(17a) $L_{n+m} + (-1)^m L_{n-m} = L_m L_n$

(17b) $L_{n+m} - (-1)^m L_{n-m} = 5F_m F_n$

(17c) $L_{2n} + (-1)^n.2 = L_n^2$

(18) $G_{n+h}H_{n+k} - G_n H_{n+h+k} = (-1)^n (G_h H_k - G_0 H_{h+k})$

(19a) $L_{n+h}F_{n+k} - L_n F_{n+h+k} = (-1)^{n+1} F_h L_k$

(19b) $F_{n+h}L_{n+k} - F_n L_{n+h+k} = (-1)^n F_h L_k$

(20a) $F_{n+h}F_{n+k} - F_n F_{n+h+k} = (-1)^n F_h F_k$

(20b) $L_{n+h}L_{n+k} - L_n L_{n+h+k} = (-1)^{n+1} 5F_h F_k$

(21) $G_m F_n - G_n F_m = (-1)^{n+1} G_0 F_{m-n} = (-1)^m G_0 F_{n-m}$

(22) $L_h^2 - 2L_{2h} = -5F_h^2$

(23) $L_{2h} - 2(-1)^h = 5F_h^2$

(24) $5F_h^2 - L_h^2 = 4(-1)^{h+1}$

(25) $5(F_h^2 + F_{h+1}^2) = L_h^2 + L_{h+1}^2 = 5F_{2h+1}$

(26) $\frac{1}{2}(3F_i + L_i) = F_{i+2}$

(27) $\frac{1}{2}(5F_i + 3L_i) = L_{i+2}$

(28) $$G_{n+1}G_{n-1} - G_n^2 = (-1)^n(G_1^2 - G_0G_2)$$

(29) $$F_{n+1}F_{n-1} - F_n^2 = (-1)^n$$

(30) $$F_{n+1}L_n = F_{2n+1} - 1 \quad \text{(for odd } n\text{)}$$

(31) $$F_{n+1}L_n = F_{2n+1} + 1 \quad \text{(for even } n\text{)}$$

(32) $$F_n^2(F_{m+1}F_{m-1}) - F_m^2(F_{n+1}F_{n-1}) = (-1)^{n-1}(F_{m+n}F_{m-n})$$

(33) $$\sum_{i=1}^{n} G_i = G_{n+2} - G_2$$

(34) $$\sum_{i=1}^{n} G_{2i-1} = G_{2n} - G_0$$

(35) $$\sum_{i=1}^{n} G_{2i} = G_{2n+1} - G_1$$

(36) $$\sum_{i=1}^{n} G_{2i} - \sum_{i=1}^{n} G_{2i-1} = G_{2n-1} + G_0 - G_1$$

(37) $$\sum_{k=1}^{n} G_{k-1}/2^k = \tfrac{1}{2}(G_0 + G_3) - G_{n+2}/2^n$$

(37a) $$\sum_{k=1}^{n} F_{k-1}/2^k = 1 - F_{n+2}/2^n$$

(38) $$\sum_{i=1}^{4k+2} G_i = L_{2k+1}G_{2k+3}$$

(39) $$\sum_{i=1}^{2n} G_iG_{i-1} = G_{2n}^2 - G_0^2$$

(40) $$\sum_{i=1}^{2n} F_iF_{i-1} = F_{2n}^2$$

(41) $$\sum_{i=1}^{2n+1} G_iG_{i-1} = G_{2n+1}^2 - G_0^2 - (G_1^2 - G_0G_2)$$

(42) $$\sum_{i=1}^{2n+1} F_i F_{i=1} = F_{2n+1}^2 - 1$$

(43) $$\sum_{i=1}^{n} G_{i+2} G_{i-1} = G_{n+1}^2 - G_1^2$$

(44) $$\sum_{i=1}^{n} G_i^2 = G_n G_{n+1} - G_0 G_1$$

(45) $$\sum_{1}^{n} F_i^2 = F_n F_{n+1}$$

(46) $$G_{n+p} = \sum_{i=0}^{p} \binom{p}{i} G_{n-i}$$

(47) $$G_{2n} = \sum_{i=0}^{n} \binom{n}{i} G_i$$

(48) $$G_{m+tp} \equiv \sum_{i=0}^{p} \binom{p}{i} G_{m+(t-2)p+i}$$

(49) $$G_{m+2p} = \sum_{i=0}^{p} \binom{p}{i} G_{m+i}$$

(50) $$F_{2n+1} = 1 + \sum_{i=0}^{n} \binom{n+1}{i+1} F_i$$

(51) $$G_{n-p} = \sum_{i=0}^{p} (-1)^i \binom{p}{i} G_{n+p-i}$$

(52) $$G_{m-tp} = \sum_{i=0}^{p} (-1)^i \binom{p}{i} G_{m-(t-1)p+i}$$

(53) $$G_{m-2p} = \sum_{i=0}^{p} (-1)^i \binom{p}{i} G_{m-p+i} = \sum_{i=0}^{p} (-1)^{p-i} \binom{p}{i} G_{m-i}$$

(54) $$F_n = \sum_{i=1}^{\infty} \binom{n-i-1}{i}$$

(55) $$G_n = \alpha\tau^n + \beta\sigma^n$$

(56) $$\alpha = (G_1 - G_0\sigma)/\sqrt{5}, \quad \beta = (G_0\tau - G_1)/\sqrt{5}$$

(57) $$\alpha\beta = (G_0G_2 - G_1^2)/5$$

(58) $$F_n = (\tau^n - \sigma^n)/\sqrt{5}$$

(59) $$L_n = \tau^n + \sigma^n$$

(60) $$\sum_{i=1}^{\infty} F_i/2^i = 2$$

(61) $$\sum_{i=1}^{\infty} iF_i/2^i = 10$$

(62) $$F_n = [\tau^n/\sqrt{5} + \tfrac{1}{2}]$$

(63) $$L_n = [\tau^n + \tfrac{1}{2}] \ (n \geqslant 2)$$

(64) $$F_{n+1} = [\tau F_n + \tfrac{1}{2}] \ (n > 1)$$

(65) $$L_{n+1} = [\tau L_n + \tfrac{1}{2}] \ (n > 3)$$

(66) $$G_{m+tp} = \sum_{i=0}^{p} \binom{p}{i} F_t^i F_{t-1}^{p-i} G_{m+i}$$

(67) $$L_{3n} = 2^n L_n + \binom{n}{1} 2^{n-1} L_{n-1} + \ldots + \binom{n}{n-1} 2L_1 + L_0$$

(68) $$F_{3n}=2^nF_n+\binom{n}{1}2^{n-1}F_{n-1}+\ldots+\binom{n}{n-1}2F_1$$

(69) $$\sum_{i=0}^{2n}\binom{2n}{i}F_{2i}=5^nF_{2n}$$

(70) $$\sum_{i=0}^{2n+1}\binom{2n+1}{i}F_{2i}=5^nL_{2n+1}$$

(71) $$\sum_{i=0}^{2n}\binom{2n}{i}L_{2i}=5^nL_{2n}$$

(72) $$\sum_{i=0}^{2n+1}\binom{2n+1}{i}L_{2i}=5^{n+1}F_{2n+1}$$

(73) $$\sum_{i=0}^{2n}\binom{2n}{i}F_i^2=5^{n-1}L_{2n}$$

(74) $$\sum_{0}^{2n+1}\binom{2n+1}{i}F_i^2=5^nF_{2n+1}$$

(75) $$\sum_{i=0}^{2n}\binom{2n}{i}L_i^2=5^nL_{2n}$$

(76) $$\sum_{i=0}^{2n+1}\binom{2n+1}{i}L_i^2=5^{n+1}F_{2n+1}$$

(77) $$\sum_{i=1}^{\infty}1/F_i=3+\sigma=4-\tau$$

(78) $$L_t^k=\sum_{i=0}^{(k-1)/2}\binom{k}{i}(-1)^{it}L_{(k-2i)t}\ \text{(for odd } k\text{)}$$

(79) $$L_t^k=\sum_{i=0}^{(k/2)-1}\binom{k}{i}(-1)^{it}L_{(k-2i)t}+\binom{k}{\frac{1}{2}k}(-1)^{tk/2}\ \text{(for even } k\text{)}$$

(80) $$(\sqrt{5})^kF_t^k=\sum_{i=0}^{(k-1)/2}\binom{k}{i}(-1)^{i(t+1)}\sqrt{5}F_{(k-2i)t}\ \text{(for odd } k\text{)}$$

(81) $$(\sqrt{5})^k F_t^k = \sum_{i=0}^{(k/2)-1} \binom{k}{i} (-1)^{i(t+1)} L_{(k-2i)t} + \binom{k}{\frac{1}{2}k} (-1)^{(t+1)k/2} \text{ (for even } k)$$

(82) $$L_{kt} = L_t^k + \sum_{i=1}^{[k/2]} \frac{k}{i} (-1)^{i(t+1)} L_t^{k-2i} \binom{k-i-1}{i-1}$$

(83) $$F_{kt} = (\sqrt{5})^{k-1} F_t^k + \sum_{i=1}^{(k-1)/2} \frac{k}{i} (-1)^{it} (\sqrt{5})^{k-2i-1} \binom{k-i-1}{i-1} F_t^{k-2i} \text{ (for odd } k)$$

(84) $$L_{kt} = (\sqrt{5})^k F_t^k + \sum_{i=1}^{k/2} \frac{k}{i} (-1)^{it} (\sqrt{5})^{k-2i} \binom{k-i-1}{i-1} F_t^{k-2i} \text{ (for even k)}$$

(85) $$\frac{F_{kt}}{F_t} = \sum_{i=0}^{(k-3)/2} (-1)^{it} L_{(k-2i-1)t} + (-1)^{(k-1)t/2} \text{ (for odd } k \geqslant 3)$$

(86) $$\frac{F_{kt}}{F_t} = \sum_{i=0}^{(k/2)-1} (-1)^{it} L_{(k-2i-1)t} \text{ (for even } k \geqslant 2)$$

(87) $$\frac{L_{kt}}{L_t} = \sum_{i=0}^{(k-3)/2} (-1)^{i(t+1)} L_{t(k-2i-1)} + (-1)^{(k-1)(t+1)/2} \text{ (for odd } k \geqslant 3)$$

(88) $$\frac{F_{kt}}{L_t} = \sum_{i=0}^{(k/2)-1} (-1)^{i(t+1)} F_{(k-2i-1)t} \text{ (for even } k \geqslant 2)$$

(89) $$\sum_{i=1}^{n} \frac{(-1)^{2^{i-1}r}}{F_{2^ir}} = \frac{(-1)^r}{F_r} \frac{F_{(2^n-1)r}}{F_{2^nr}}$$

(90) $$-1 + 1/F_4 + 1/F_8 + \ldots + 1/F_{2^n} = -F_{2^n-1}/F_{2^n}$$

(91) $$F_n = \frac{1}{2^{n-1}} \left[\binom{n}{1} + 5\binom{n}{3} + 5^2\binom{n}{5} + \ldots \right]$$

(92) $$L_n = \frac{1}{2^{n-1}} \left[1 + 5\binom{n}{2} + 5^2\binom{n}{4} + \ldots \right]$$

(93) $$5[F_r^2(-1)^r + F_{2r}^2 + F_{3r}^2(-1)^{3r} + \ldots + F_{(n+1)r}^2(-1)^{(n+1)r}] = = (-1)^{(n+1)r}F_{(2n+3)r}/F_r - 2n - 3$$

(94) $$L_r^2(-1)^r + L_{2r}^2(-1)^{2r} + \ldots + L_{(n+1)r}^2(-1)^{(n+1)r} = = (-1)^{(n+1)r}F_{(2n+3)r}/F_r + 2n + 1$$

(95) $$5[F_1^2 + F_3^2 + \ldots + F_{2n-1}^2] = F_{4n} + 2n$$

(96) $$L_1^2 + L_3^2 + \ldots + L_{2n-1}^2 = F_{4n} - 2n$$

(97) $$\sum_{i=0}^{n}(-1)^i L_{n-2i} = 2F_{n+1}$$

(98) $$5\sum_{i=0}^{n}F_iF_{n-i} = (n+1)L_n - 2F_{n+1} = nL_n - F_n$$

(99) $$\sum_{i=0}^{n}L_iL_{n-1} = (n+1)L_n + 2F_{n+1} = (n+2)L_n + F_n$$

(100) $$\sum_{r=0}^{n}F_rL_{n-r} = (n+1)F_n$$

(101) $$\lim_{n=\infty} F_{n+1}/F_n = \tau$$

(101a) $$\lim_{n=\infty} F_n/F_{n-m} = \tau^m$$

(102) $$1 + \sum_{n=2}^{\infty}\frac{(-1)^n}{F_nF_{n-1}} = \tau$$

(103a) $F_{n+1}/F_n < F_{n-1}/F_{n-2}$ when n is even

$F_{n+1}/F_n > F_{n-1}/F_{n-2}$ when n is odd.

(103b) $$F_{n+1} - F_n\tau = \frac{(-1)^n}{F_{n-1} + F_n\tau}$$

(103c) $$\left|\tau - F_{n+1}/F_n\right| < \left|\tau - F_n/F_{n-1}\right|$$

(104) $$\left|\tau - \mathrm{F}_{\mathrm{n}+1}/\mathrm{F}_{\mathrm{n}}\right| < \frac{1}{\mathrm{F}_{\mathrm{n}}^2}$$

(105) $$\tau = \prod_{i=1}^{\infty}\left(1 + \frac{(-1)^{i+1}}{F_{i+1}^2}\right)$$

(106) $$\frac{F_{(t+1)m}}{F_{tm}} = L_m - \cfrac{(-1)^m}{L_m - \cfrac{(-1)^m}{L_m - \cdots \cfrac{}{\cfrac{(-1)^m}{L_m}}}}$$

References

Ball, W. W. R. and Coxeter, H.S. M. (1947). *Mathematical recreations and essays.* (Macmillan).

Bellman, R. E. and Dreyfus, S. E. (1962). *Applied dynamic programming,* pp. 152–155 (Princeton U.P.)

Binet, J. P. M. (1843). *C. R. Paris* **18**, 563; (1844) **19**, 939.

Bowcamp, C. J. (1965). Solution to problem 63–14, a resistance problem. *S.I.A.M. Review* **7**, 286–290.

Carlitz, L. (1968). Fibonacci representation. *Fib. Quart.* **6**, 193–220.

Carlitz, L. and Ferns, H. H. (1970). Some Fibonacci and Lucas identities. *Fib. Quart.* **8**, 61–73.

Carlitz, L., Scoville, R. and Hoggatt jr, V. E. (1972a). Fibonacci representation. *Fib. Quart.* **10**, 1–28; addition (1973) **11**, 527–530.

Carlitz, L., Scoville, R. and Hoggatt jr, V. E. (1972b). Lucas representation. *Fib. Quart.* **10**, 29–42, 70, 112.

Carmichael, R. D. (1913/14). On the numerical factors of the arithmetic forms $\alpha^n \pm \beta^n$. *Ann. of Math.* **15** (2), 30–70.

Catalan, E. (1857). *Manuel des candidats à l'École Polytechnique*, I.86.

Catalan, E. (1886). *Mém. Soc. Roy. Sc. Liège* **13** (2), 319–321 (*Mélanges Math.* II).

Cavachi, M. (1980). Unele proprietaţi de termenilor şirului lui Fibonacci. *Gazeta Matem.* **85**, 290–293.

Chang, D. K. (1986). Higher-order Fibonacci sequences modulo M. *Fib. Quart.* **24**, 138–139.

Conolly, B. W. (1981). *Techniques of operational research.* (Ellis Horwood)

Conolly, B. W. (1988). Private Communication.

Danese, A. E. (1960). Problems for solution. *Amer. Math. Mthly* **67**, 8.

Dickson, L. E. (1952). *History of the theory of numbers I.* Chapter XVII. (Chelsea P.C.)

d'Ocagne, M. (1885/86). *Bull. Soc. Math. France* **14**, 20–41.

Euler, L. (1737). De fractionibus continuis observationes. *C. Petr.* **9**.

Forfar, D. O. and Keogh, T. W. (1985/86). Independence and the length of a run of wins. *Math. Spectrum* **18**, 70–75.

Gardner, M. (1956). *Mathematics, magic and mystery*. (Dover)
Gardner, M. (1968). *Mathematical circus*. (Penguin)
Good, I. J. (1974). A reciprocal series of Fibonacci numbers. *Fib. Quart.* **12**, 346.
Gordon, J. (1934). *A step-ladder to painting*, 143 pp. (Faber)
Hardy, G. H. and Wright, E. M. (1980). *An introduction to the theory of numbers*. 5th ed. (Oxford U.P.)
Hoggatt jr, V. E. (1969). *Fibonacci and Lucas numbers*. (Houghton-Mifflin)
Hoggatt jr, V. E. and Bicknell, M. (1974). Some congruences of the Fibonacci numbers modulo a prime p. *Math. Mag.* **47**, 210–214.
Holden, H. (1975). Fibonacci Tiling. *Fib. Quart.* **13**, 45–49.
Horadam, A. F. (1961). A generalized Fibonacci sequence. *Amer. Math. Mthly* **68**, 455–459.
Horadam, A. F. (1965). Generating functions for powers of a certain generalized sequence of numbers. *Duke Math. J.* **32**, 437–446.
Hounslow, G. (1973). Phillotaxis, the arrangement of leaves according to mathematical laws. *Ill. Vict. Inst.* **VI**, 129–140.
Hunter, J. A. H. (1963). *Fib. Quart.* **1**, 66.
Hunter, J. A. H. (1964). *Fib. Quart.* **2**, 104.
Huntley, H. E. (1970). *The divine proportion. A study in mathematical beauty*. (Dover)
Isaacs, R. P. (1958). Mentioned in Gardner, M. (1977). Mathematical Games. Scientific American.
Jarden, D. (1946). Two theorems on Fibonacci's sequence. *Amer. Math. Mthly* **53**, 425–427.
Jones, J. P. (1975). Diophantine representation of the Fibonacci numbers. *Fib. Quart.* **13**, 84–88.
Jones, J. P. (1976). Diophantine representation of the Lucas numbers. *Fib. Quart.* **14**, 134.
Kenyon, J. C. (1967). Nim-like games and the Sprague–Grundy theory. Thesis, Univ. Calgary, Alberta.
Kiefer, J. (1953). Sequential minimax search for a maximum. *Proc. Amer. Math. Soc.* **4**, 502–506.
Kramer, J. and Hoggatt jr, V. E. (1973). Special cases of Fibonacci periodicity. *Fib. Quart.* **11**, 519–522.
Kuipers, L. and Shiua, J.-S. (1972). A distribution property of the sequence of Fibonacci numbers. *Fib. Quart.* **10**, 375–376, 392.
Lamé, G. (1844). *C. R. Paris* **19**, 867.
Lekkerkerker, C. G. (1952). Voorstelling van natuurlijke getallen door een som van getallen van Fibonacci. *Simon Stevin* **29**, 190–195.
Lloyd, E. K. (1985). Enumeration. *Handbook of applicable mathematics*, Chap. 15, Vol. V. (Wiley)
Lucas, E. (1878). Theorie des fonctions numériques simplement périodiques. *Amer. J. Math.* **I**, 184–240, 289–321.
Miles jr, E. P. (1960). Generalised Fibonacci numbers and associated matrices. *Amer. Math. Mthly* **67**, 745–752.
Myers, B. R. (1971). Number of spanning trees in a wheel. *IERR Trans. Circuit Theory CT*-18, 280–281.

Miller, J. C. P. and Prentice, M. J. (1968). *Computer J.* **11**, 241–246.
Narkiewicz, W. (1984). *Uniform distribution of sequence of integers in residue-classes.* (Lecture notes in mathematics 1087) (Springer).
Netto, E. (1901). *Lehrbuch der Kombinatorik* (Teubner).
Niederreiter, H. (1972). Distribution of Fibonacci numbers modulo 5^k. *Fib. Quart.* **10**, 373–374.
Owings jr, J. C. (1987). Solution of the system $a^2 \equiv -1(\text{mod}b)$, $b^2 \equiv -1(\text{mod}a)$. *Fib. Quart.* **25**, 245–249.
Pacioli, L. (1509). *De divina proportione.*
Parman and Singh (1985). *Historia matematica* **12**, 229–244.
Perron, O. (1929). *Die Lehre von den Kettenbrüchen*, 2nd ed. (Chelsea)
Phillips, G. M. (1984). Aitken sequences and Fibonacci numbers. *Amer. Math. Mthly* **91**, 354–357.
Rényi, A. (1984). Variations on a theme by Fibonacci. In *A diary on information theory*. (Akad. kiadó, Budapest and Wiley)
Riordan, J. (1962). Generating functions for powers of Fibonacci numbers. *Duke Math. J.* **29**, 5–12.
Schönemann (1873). Über die Konstruktion und Darstellung des Ikosaeders und Sternikosaeders. *Zeitschr. f. Math. und Phys.* **18**, 387–392.
Sedlaček, J. (1969). On the number of spanning trees of finite graphs. *Časopis propěstovani matematiky* **94**, 217–221.
Siebeck, H. (1846). *J. f. Math.* **33**, 71–76.
Singmaster, D. (1975). Repeated binomial coefficients and Fibonacci numbers. *Fib. Quart.* **13**, 295–298.
Świerczkowski, S. (1958). On successive settings of an arc on the circumference of a circle. *Fund. mat.* **46**, 187–189.
Thompson, D'A. W. (1963). *On growth and form*, Vol. II. (Cambridge U.P.)
Tucker, A. (1980) *Applied combinatorics*. (Wiley)
Turner, J. C. (1986). On caterpillars, trees and stochastic processes. *Amer. Math. Mthly* **93**, 205, 213.
Wall, D. D. (1960). Fibonacci series modulo *m*. *Amer. Math. Mthly* **67**, 525–532.
Watson, F. R. (1983). Investigating some number sequences. *Bull. I.M.A.* **19**, 149–153.
Whinihan, M. J. (1963). Fibonacci NIM. *Fib. Quart.* **1**, 9–13.
Wythoff, W. A. (1907). A modification of the game of NIM. *Nieuw Archief voor wiskunde* **7** (2), 199.

Fibonacci and Lucas numbers and their prime factors

F_n	n	L_n
1	1	1
1	2	3
2	3	$4 = 2^2$
3	4	7
5	5	11
$8 = 2^3$	6	$18 = 2\times3^2$
13	7	29
$21 = 3\times7$	8	47
$34 = 2\times17$	9	$76 = 2^2\times19$
$55 = 5\times11$	10	$123 = 3\times41$
89	11	199
$144 = 2^4\times3^2$	12	$322 = 2\times7\times23$
233	13	521
$377 = 13\times29$	14	$843 = 3\times281$
$610 = 2\times5\times61$	15	$1364 = 2^2\times11\times31$
$987 = 3\times7\times47$	16	2207
1597	17	3571
$2584 = 2^3\times17\times19$	18	$5778 = 2\times3^3\times107$
$4181 = 37\times113$	19	9349
$6765 = 3\times5\times11\times41$	20	$15127 = 7\times2161$
$10946 = 2\times13\times421$	21	$24476 = 2^2\times29\times211$
$17711 = 89\times199$	22	$39603 = 3\times43\times307$
28657	23	$64079 = 139\times461$
$46368 = 2^5\times3^2\times7\times23$	24	$103682 = 2\times47\times1103$
$75025 = 5^2\times3001$	25	$167761 = 11\times101\times151$
$121393 = 233\times521$	26	$271443 = 3\times90481$

F_n	n	L_n
$196418 = 2\times17\times53\times109$	27	$439204 = 2^2\times19\times5779$
$317811 = 3\times13\times29\times281$	28	$710647 = 7^2\times14503$
514229	29	$1149851 = 59\times19489$
$832040 = 2^3\times5\times11\times31\times61$	30	$1860498 = 2\times3^2\times41\times2521$

Index

MATHEMATICS AND ITS APPLICATIONS

Series Editor: G. M. BELL, Professor of Mathematics,
King's College London (KQC), University of London

NUMERICAL ANALYSIS, STATISTICS AND OPERATIONAL RESEARCH

Editor: B. W. CONOLLY, Emeritus Professor of Mathematics (Operational Research),
Queen Mary College, University of London

Mathematics and its applications are now awe-inspiring in their scope, variety and depth. Not only is there rapid growth in pure mathematics and its applications to the traditional fields of the physical sciences, engineering and statistics, but new fields of application are emerging in biology, ecology and social organization. The user of mathematics must assimilate subtle new techniques and also learn to handle the great power of the computer efficiently and economically.

The need for clear, concise and authoritative texts is thus greater than ever and our series will endeavour to supply this need. It aims to be comprehensive and yet flexible. Works surveying recent research will introduce new areas and up-to-date mathematical methods. Undergraduate texts on established topics will stimulate student interest by including applications relevant at the present day. The series will also include selected volumes of lecture notes which will enable certain important topics to be presented earlier than would otherwise be possible.

In all these ways it is hoped to render a valuable service to those who learn, teach, develop and use mathematics.

Mathematics and its Applications

Series Editor: G. M. BELL, Professor of Mathematics, King's College London (KQC), University of London

Anderson, I.	**Combinatorial Designs**
Artmann, B.	**Concept of Number: From Quaternions to Monads and Topological Fields**
Arczewski, K. & Pietrucha, J.	**Mathematical Modelling in Discrete Mechanical Systems**
Arczewski, K. and Pietrucha, J.	**Mathematical Modelling in Continuous Mechanical Systems**
Bainov, D.D. & Konstantinov, M.	**The Averaging Method and its Applications**
Baker, A.C. & Porteous, H.L.	**Linear Algebra and Differential Equations**
Balcerzyk, S. & Jösefiak, T.	**Commutative Rings**
Balcerzyk, S. & Jösefiak, T.	**Commutative Noetherian and Krull Rings**
Baldock, G.R. & Bridgeman, T.	**Mathematical Theory of Wave Motion**
Ball, M.A.	**Mathematics in the Social and Life Sciences: Theories, Models and Methods**
de Barra, G.	**Measure Theory and Integration**
Bartak, J., Herrmann, L., Lovicar, V. & Vejvoda, D.	**Partial Differential Equations of Evolution**
Bell, G.M. and Lavis, D.A.	**Co-operative Phenomena in Lattice Models, Vols. I & II**
Berry, J.S., Burghes, D.N., Huntley, I.D., James, D.J.G. & Moscardini, A.O.	**Mathematical Modelling Courses**
Berry, J.S., Burghes, D.N., Huntley, I.D., James, D.J.G. & Moscardini, A.O.	**Mathematical Modelling Methodology, Models and Micros**
Berry, J.S., Burghes, D.N., Huntley, I.D., James, D.J.G. & Moscardini, A.O.	**Teaching and Applying Mathematical Modelling**
Blum, W.	**Applications and Modelling in Learning and Teaching Mathematics**
Brown, R.	**Topology: A Geometric Account of General Topology, Homotopy Types and the Fundamental Groupoid**
Burghes, D.N. & Borrie, M.	**Modelling with Differential Equations**
Burghes, D.N. & Downs, A.M.	**Modern Introduction to Classical Mechanics and Control**
Burghes, D.N. & Graham, A.	**Introduction to Control Theory, including Optimal Control**
Burghes, D.N., Huntley, I. & McDonald, J.	**Applying Mathematics**
Burghes, D.N. & Wood, A.D.	**Mathematical Models in the Social, Management and Life Sciences**
Butkovskiy, A.G.	**Green's Functions and Transfer Functions Handbook**
Cartwright, M.	**Fourier Methods: Applications in Mathematics, Engineering and Science**
Cerny, I.	**Complex Domain Analysis**
Chorlton, F.	**Textbook of Dynamics, 2nd Edition**
Chorlton, F.	**Vector and Tensor Methods**
Cohen, D.E.	**Computability and Logic**
Cordier, J.-M. & Porter, T.	**Shape Theory: Categorical Methods of Approximation**
Crapper, G.D.	**Introduction to Water Waves**
Cross, M. & Moscardini, A.O.	**Learning the Art of Mathematical Modelling**
Cullen, M.R.	**Linear Models in Biology**
Dunning-Davies, J.	**Mathematical Methods for Mathematicians, Physical Scientists and Engineers**
Eason, G., Coles, C.W. & Gettinby, G.	**Mathematics and Statistics for the Biosciences**
El Jai, A. & Pritchard, A.J.	**Sensors and Controls in the Analysis of Distributed Systems**
Exton, H.	**Multiple Hypergeometric Functions and Applications**
Exton, H.	**Handbook of Hypergeometric Integrals**
Exton, H.	***q*-Hypergeometric Functions and Applications**
Faux, I.D. & Pratt, M.J.	**Computational Geometry for Design and Manufacture**
Firby, P.A. & Gardiner, C.F.	**Surface Topology**
Gardiner, C.F.	**Modern Algebra**

Mathematics and its Applications

Series Editor: G. M. BELL, Professor of Mathematics, King's College London (KQC), University of London

Gardiner, C.F.	**Algebraic Structures**
Gasson, P.C.	**Geometry of Spatial Forms**
Goodbody, A.M.	**Cartesian Tensors**
Goult, R.J.	**Applied Linear Algebra**
Graham, A.	**Kronecker Products and Matrix Calculus: with Applications**
Graham, A.	**Matrix Theory and Applications for Engineers and Mathematicians**
Graham, A.	**Nonnegative Matrices and Applicable Topics in Linear Algebra**
Griffel, D.H.	**Applied Functional Analysis**
Griffel, D.H.	**Linear Algebra and its Applications: Vol 1, A First Course; Vol. 2, More Advanced**
Guest, P. B.	**The Laplace Transform and Applications**
Hanyga, A.	**Mathematical Theory of Non-linear Elasticity**
Harris, D.J.	**Mathematics for Business, Management and Economics**
Hart, D. & Croft, A.	**Modelling with Projectiles**
Hoskins, R.F.	**Generalised Functions**
Hoskins, R.F.	**Standard and Non-standard Analysis**
Hunter, S.C.	**Mechanics of Continuous Media, 2nd (Revised) Edition**
Huntley, I. & Johnson, R.M.	**Linear and Nonlinear Differential Equations**
Irons, B. M. & Shrive, N. G.	**Numerical Methods in Engineering and Applied Science**
Ivanov, L. L.	**Algebraic Recursion Theory**
Johnson, R.M.	**Theory and Applications of Linear Differential and Difference Equations**
Johnson, R.M.	**Calculus: Theory and Applications in Technology and the Physical and Life Sciences**
Jones, R.H. & Steele, N.C.	**Mathematics in Communication Theory**
Jordan, D.	**Geometric Topology**
Kelly, J.C.	**Abstract Algebra**
Kim, K.H. & Roush, F.W.	**Applied Abstract Algebra**
Kim, K.H. & Roush, F.W.	**Team Theory**
Kosinski, W.	**Field Singularities and Wave Analysis in Continuum Mechanics**
Krishnamurthy, V.	**Combinatorics: Theory and Applications**
Lindfield, G. & Penny, J.E.T.	**Microcomputers in Numerical Analysis**
Livesley, K.	**Engineering Mathematics**
Lootsma, F.	**Operational Research in Long Term Planning**
Lord, E.A. & Wilson, C.B.	**The Mathematical Description of Shape and Form**
Malik, M., Riznichenko, G.Y. & Rubin, A.B.	**Biological Electron Transport Processes and their Computer Simulation**
Massey, B.S.	**Measures in Science and Engineering**
Meek, B.L. & Fairthorne, S.	**Using Computers**
Menell, A. & Bazin, M.	**Mathematics for the Biosciences**
Mikolas, M.	**Real Functions and Orthogonal Series**
Moore, R.	**Computational Functional Analysis**
Murphy, J.A., Ridout, D. & McShane, B.	**Numerical Analysis, Algorithms and Computation**
Nonweiler, T.R.F.	**Computational Mathematics: An Introduction to Numerical Approximation**
Ogden, R.W.	**Non-linear Elastic Deformations**
Oldknow, A.	**Microcomputers in Geometry**
Oldknow, A. & Smith, D.	**Learning Mathematics with Micros**
O'Neill, M.E. & Chorlton, F.	**Ideal and Incompressible Fluid Dynamics**
O'Neill, M.E. & Chorlton, F.	**Viscous and Compressible Fluid Dynamics**
Page, S. G.	**Mathematics: A Second Start**
Prior, D. & Moscardini, A.O.	**Model Formulation Analysis**
Rankin, R.A.	**Modular Forms**
Scorer, R.S.	**Environmental Aerodynamics**
Shivamoggi, B.K.	**Stability of Parallel Gas Flows**
Smith, D.K.	**Network Optimisation Practice: A Computational Guide**
Srivastava, H.M. & Manocha, L.	**A Treatise on Generating Functions**
Stirling, D.S.G.	**Mathematical Analysis**
Sweet, M.V.	**Algebra, Geometry and Trigonometry in Science, Engineering and Mathematics**
Temperley, H.N.V.	**Graph Theory and Applications**
Temperley, H.N.V.	**Liquids and Their Properties**
Thom, R.	**Mathematical Models of Morphogenesis**
Thurston, E.A.	**Techniques of Primary Mathematics**
Toth, G.	**Harmonic and Minimal Maps and Applications in Geometry and Physics**
Townend, M. S.	**Mathematics in Sport**
Townend, M.S. & Pountney, D.C.	**Computer-aided Engineering Mathematics**
Trinajstic, N.	**Mathematical and Computational Concepts in Chemistry**
Twizell, E.H.	**Computational Methods for Partial Differential Equations**
Twizell, E.H.	**Numerical Methods, with Applications in the Biomedical Sciences**
Vince, A. and Morris, C.	**Mathematics for Information Technology**
Walton, K., Marshall, J., Gorecki, H. & Korytowski, A.	**Control Theory for Time Delay Systems**
Warren, M.D.	**Flow Modelling in Industrial Processes**
Wheeler, R.F.	**Rethinking Mathematical Concepts**
Willmore, T.J.	**Total Curvature in Riemannian Geometry**
Willmore, T.J. & Hitchin, N.	**Global Riemannian Geometry**